D1250145

A MATHEMATICAL MOSAIC

A MATHEMATICAL MOSAIC

Patterns & Problem Solving

by

Ravi Vakil

Brendan Kelly Publishing Inc.
2122 Highview Drive
Burlington, Ontario L7R 3X4

Cover design: Pronk&Associates
Toronto, Ontario

Illustrations: Taisa Kelly

Brendan Kelly Publishing Inc.
2122 Highview Drive
Burlington, Ontario L7R 3X4

ISBN 1-895997-04-6

ATTENTION EDUCATIONAL ORGANIZATIONS

Quantity discounts are available on bulk purchases of this book for educational
purposes or fund raising. For information, please contact:

Brendan Kelly Publishing Inc.
2122 Highview Drive
Burlington, Ontario
L7R 3X4

Telephone: (905) 335-5954   Fax: (905) 335-5104

to the memory of my father

Acknowledgments

Mathematics is an intensely social discipline. My first debt is to those who have shared their love of mathematics with me: my teachers, professors, colleagues, students, and friends. I am especially grateful to the young men and women who agreed to share their unique perspectives on the joys of mathematics for the seven profiles included in this book. Most of the ideas contained within *A Mathematical Mosaic* have been passed on to me over the years by word of mouth. Although they are presented here in new and, I hope, engaging ways, the origins of many of these gems have been lost in mathematical folklore. I have cited references where I am aware of them, but there are many other sources that I must gratefully, if silently, acknowledge here.

Michael Roth, a graduate student in mathematics at Harvard University, and Naoki Sato, a talented undergraduate at the University of Toronto, both gave essential advice on the final draft. Mark Wunderlich, formerly a student of mine at Harvard College and now a graduate student in philosophy at the University of Arizona, provided feedback during the development of much of this material. The book is far more readable thanks to their efforts; any flaws that remain are solely the responsibility of the author.

From helping to develop the original concept to suggesting final revisions, Brendan Kelly transcended his role as publisher. This book has evolved during long conversations with him, and his influence is visible on every page. I greatly appreciate his creative vision and commitment to this rather unorthodox project.

Finally, I must thank Alice Staveley of Oxford University for her meticulous editorial advice and endless patience.

Publisher's Introduction

A Celebration of the Human Intellect

A well-known story from the folklore of mathematical history tells how Archimedes bounded from his bathtub and ran naked through the streets of ancient Greece yelling, "Eureka!" [I have found it]. Apparently a period of intense meditation in the bathtub had provided the inspiration for Archimedes' discovery of the law of hydrostatics which now bears his name. Whether apocryphal or authentic, this legend captures the ecstasy which accompanies a flash of insight into a deep problem. It is to those who relish the joy of mathematical discovery that this book is dedicated. It is for those who chase relentlessly the thrill of *eureka* that this book is written.

Through the past three millennia, the mental giants of each generation have pitted their problem solving skills against the most intractable problems to challenge the human mind. When a problem yielded like a dragon to a deft sword, the successful strategy was refined, polished, and added to the growing body of intellectual technique known as *mathematics*. One of the greatest problem solvers of all time, Isaac Newton, characterized his favorite discoveries as "smoother pebbles or prettier shells than ordinary".[1] From an assortment of the most beautiful pebbles and the prettiest shells of mathematical technique, Ravi Vakil, author of this book, has created an intriguing mosaic. It is a mosaic which links various branches of mathematics through powerful overarching ideas. Ravi himself is a preeminent problem solver, having won top honors on virtually every mathematics competition and olympiad. His remarkable achievements are outlined on page 9 and on the back cover.

Throughout this book, Ravi Vakil profiles several other recent winners of mathematical competitions and olympiads: young people who have honed their intellectual gifts to world class levels. As you read these human interest features, you will observe that these fine young minds all share a passion for the eureka sensation. It is something they all understand and something that bonds them together in friendly collegial competitions.

We are pleased to publish this celebration of the human intellect. It is offered as a salute to the problem solvers of the present and future and as a tribute to the intellectual giants of the past. To the reader, we issue a challenge. We dare you to flip through the pages of this book without finding any problems to pique your curiosity!

---

[1]For the complete quote, see page 8.

I do not know what I may appear to the world; but to myself I seem to have been only like a boy playing on the seashore, and diverting myself in now and then finding a smoother pebble or a prettier shell than ordinary, whilst the great ocean of truth lay all undiscovered before me.

— Isaac Newton

About the Author

PHOTO BY JEWEL RANDOLPH

Ravi Vakil's resumé reads like a dictionary of superlatives. During his high school years he won every major mathematics competition. These include first place standing in the *Canadian Mathematical Olympiad* for two successive years, first place standing in North America on the 1988 *USA Mathematical Olympiad,* and two gold medals and one silver medal in the *International Mathematical Olympiad.* He also won first prize in the *Canadian Association of Physicists Competition* and led his high school computer team to three provincial championships. Ravi was valedictorian of his 1988 graduating class at Martingrove Collegiate Institute in Metropolitan Toronto and was awarded the *Academic Gold Medal* and the *Governor General's Award* for excellence.

During his undergraduate years at the University of Toronto, Ravi raised his brilliant achievements to new heights. In all four years he placed among the top five competitors in the prestigious North American *Putnam Mathematical Competition,* qualifying him as Putnam Fellow in each year. He graduated in 1992 with B. Sc. and M. Sc. degrees; for his B. Sc., he was awarded the *Governor General's Medal* for the highest graduating marks at the University of Toronto.

Ravi's extraordinary achievements have not prevented him from pursuing his many other interests. These include squash, debating, student government, journalism, and *Amnesty International.* For this breadth of involvement he has received numerous scholarships including the *John H. Moss Memorial Fund Scholarship* for the best all-round graduating student at the University of Toronto.

In spite of these personal successes, Ravi Vakil is a compassionate person with an eager willingness to help others. He has worked extensively as a coach of the Canadian Team to the *International Mathematical Olympiad* from 1989 through 1995. He is also co-founder of *Mathematical Mayhem,* a mathematical problem-solving journal for high school and undergraduate students — the only student-run journal of its kind in the English-speaking world!

Ravi Vakil is currently a Ph.D. candidate in pure mathematics at Harvard University, working in Algebraic Geometry under the supervision of Dr. Joseph Harris.

Foreword

The public face of mathematics is sometimes grey and utilitarian: math is a useful tool, a haphazard collection of recipes and algorithms, a necessary prerequisite to understanding science. The private face is much more beautiful: Mathematics as Queen, not servant, of Science. Math is a uniquely aesthetic discipline; mathematicians use words like beauty, depth, elegance, and power to describe excellent ideas. In order to truly enjoy mathematics, one must learn to appreciate the beauty of elegant arguments, and then learn to construct them.

In *A Mathematical Mosaic,* I hope to get across a little of what it is that mathematicians actually do. Most "big ideas" and recurring themes in mathematics come up in surprisingly simple problems or puzzles that are accessible with relatively little background. Many of them have been collected here, along with an inkling of how they relate to the frontiers of modern research. These themes will also be traced backwards; often ideas centuries old take on new meaning and relevance as mathematical understanding advances. And if in the process of looking at these important ideas we get a little playful and irreverent, well, that is also in keeping with the nature of mathematics.

This book does not purport to teach problem solving, although it might communicate what is interesting and exciting about grappling with a problem. Instead, the reader should be left with a large chunk of mathematics to digest slowly.

This is not a book intended to ever be "finished with" or completed. Instead, it should be read bit by bit. Like all mathematics, it should be read with pencil in hand, with more time spent in deep thought or frantic scribbling than in actual reading. Mathematics is not a spectator sport!

You will also notice that, like any mosaic, this book comes in many small pieces of different sorts. It is loosely arranged in order of difficulty, and the headers often indicate the general theme of the article. You might enjoy just starting with one section which especially interests you; from there, you can skip to other sections which link with your earlier choices. You will soon discover that ideas from one field carry over into many others. And if you don't end up at a section that amazes and perplexes you, I will be sorely disappointed.

Many sections end with some *Food for Thought*. More than just exercises, these excursions are intended to be starting points for independent thinking. The material beforehand will help, but often only tangentially. I have deliberately included very few solutions. Indeed, often there is no single "correct" solution, and sometimes the problems are so open-ended that there is no way to completely "finish" them. Don't try to solve every one; choose one or two that catch your eye and think about them on and off for a few days. Success should not be defined by how many problems you solve, but by how many new ideas you have. And remember: patience is a virtue.

Every so often throughout the book, references are mentioned. This is not to intimidate you into reading many weighty tomes. Instead, they are there to provide some ideas for further reading in case you find some field particularly interesting.

I have also included short profiles of interesting and talented young people who have been involved in mathematics for many years. I have chosen them fairly randomly from the large number of fascinating people I have had the good fortune to meet through my involvement in mathematics education. Some of them are going on in mathematics (and indeed one has already made his name in the field), while others are pursuing their interests in related subjects. Through these profiles, I hope to share their unique perspectives on the joys of mathematics.

You will notice that these young people have all done extremely well in mathematical competitions; collectively they have won eleven Gold Medals at International Mathematical Olympiads. In noting their achievements, I certainly do not mean to imply that competitions are the only gateway to mathematics, or that they are a necessary prerequisite to success in the field. My selection merely reflects the fact that my own involvement with young mathematicians has grown out of my association with mathematics competitions.

There is no background required to begin enjoying this book. You should be able to pick up many of the ideas as you go along. When you truly get stuck, ask a friend or a teacher for help. You might be surprised at how much you learn! For some of the later sections, a familiarity with more advanced ideas such as mathematical induction, indirect arguments, set theory, complex numbers, and even calculus will be useful or even necessary, but the vast majority can be read with nothing more than early high school mathematics and a little chutzpah.

But enough of this talk. On to the mathematics!

Cambridge, Mass.
January, 1996

TABLE OF CONTENTS: Part I

Number Theory

Combinatorics

The Fibonacci Sequence

Game Theory

TABLE OF CONTENTS: *Part I*

Geometry

Combinatorics Revisited

Chessboard Coloring

TABLE OF CONTENTS: Part II

Number Theory Revisited

Fibonacci & The Golden Mean

Geometry Revisited

Infinity

TABLE OF CONTENTS: Part II

Game Theory Revisited

Concepts in Calculus

Complex Numbers

Infinity Revisited

NUMBER THEORY

Mathematics is the Queen of the Sciences, and
Number Theory the Queen of Mathematics.

— Karl Gauss

Number Theory

Calculating Tricks of the Trade

There are many "cute" little numerical tricks for performing calculations quickly. On the surface, they may not seem to be serious mathematics, but they are good practice for elementary algebra. For example, here's a little trick to multiply two positive integers with the same tens digits and with the units digits summing to ten. Study these examples to see if you can find a pattern in the answers and discover the trick.

38	13	79	25	118
× 32	× 17	× 71	× 25	× 112
1216	221	5609	625	13216

The Trick

•Multiply the tens digit by the next largest number. Call your result "A".

•Then multiply the units digits together, and call your result "B".

•Write A immediately followed by B (where B is considered a two-digit number), and read it as a single number.

•You have your product!

For example, to multiply 32 and 38 in your head, you get A = 12 and B = 16, so 32 × 38 = 1216. Use the trick to verify the other answers above. (The last example may not appear to follow the rule, but you can easily adapt it.)

The Proof

The proof isn't too difficult. (You might want to derive it yourself.)
Let x be the tens digit of both numbers, and y be the units digit of the first number.
Then $10-y$ is the units digit of the second number.
The two numbers are therefore $10x + y$ and $10x + 10-y$.
Multiplying them together, we get:

$$(10x+y)(10x + 10 -y)= 100x^2 + 100x + y(10-y) \qquad \text{Some steps have}$$
$$= 100\ x(x+1) + y(10-y) \qquad \text{been omitted.}$$
$$= 100\underset{\uparrow}{A}\ +\ B\ \leftarrow \text{product of units digits}$$

product of the number left when units digit is removed, and its successor

18

Number Theory

Mission Impossible

Here is another way to multiply numbers quickly on paper. This one will be trickier; I am not going to tell you how it works. Your mission, should you choose to accept it, is to piece together the method from only four examples. (If you can figure out *why* it works, you will have a good understanding of how computers multiply numbers.)

Example 1 $12 \times 23 = 276$

12	23
6	46
3	92
1	184
	276

Example 2 $35 \times 42 = 1470$

35	42
17	84
8	168
4	336
2	672
1	1344
	1470

Example 3 $53 \times 65 = 3445$

53	65
26	130
13	260
6	520
3	1040
1	2080
	3445

Example 4 $21 \times 71 = 1491$

21	71
10	142
5	284
2	568
1	1136
	1491

Here is another little numerical puzzle for you to ponder.

> Find a six-digit number N such that 2N, 3N, 4N, 5N, and 6N have exactly the same digits as N, but in a different order. (This number might look familiar to you.) What is 7N?

If you're stuck, the answer to this puzzle is on the next page, but it is encoded to prevent you from peeking too quickly.

Number Theory

Calculating Prodigies and Idiot Savants

Throughout history there have been people who have demonstrated super-human calculating skills. These gifted people are usually unable to explain how they perform their complex computations with computer-like speed. Since such people are not usually distinguished by exceptional mathematical ability, they have been referred to as *idiot savants*. In 1988, actor Dustin Hoffman won the best actor Academy Award for his portrayal of an autistic idiot savant in the movie, *Rain Man*. One of the most famous idiot savants in history was George Parker Bidder who at age 10 in 1816 was asked, "If a coach wheel is 5 feet 10 inches in circumference, how many times will it revolve in running 800 000 000 miles?" Bidder gave the correct answer, 724 114 285 714 times with 20 inches remaining — in 50 seconds!

More recently, the Guinness Book of Records reported that on June 18, 1980, Mrs. Shakuntala Devi of India correctly calculated the product of two randomly selected 13-digit numbers…in 28 seconds!

On April 7, 1981, Willem Klein successfully extracted the 13th root of a 100 digit number in 1 minute, 28.8 seconds at the National Laboratory for High Energy Physics in Tsukuba, Japan. In light of these remarkable feats it seems more appropriate to refer to idiot savants as calculating prodigies.

Here's the answer to the puzzle at the bottom of the previous page. Decode it by replacing each letter with the previous one in the alphabet.

UBLF UIF GJSTU TJY EJHJUT BGUFS UIF EFDJNBM QPJOU JO UIF EFDJNBM FYQBOTJPO PG POF TFWFOUI

CHALLENGE
This works in base 10. Can you think of similar examples in other bases?

Number Theory

The Magic Birthday Predictor Cards

Merlin the mathematician claims he can tell you the day of the month in which you were born, merely by asking you to identify which of five magic cards contain your date of birth. Here's how he does it.

Merlin shows these 5 cards to you and asks that you select those on which the day of your birth appears. He then adds together the numbers in the upper left corner of the cards which you have selected. This sum is the date of your birth.

Example: Suppose your birthday falls on the 12th of the month. You would select card 2 and card 3 since these are the only cards on which the number 12 appears.

4 + 8 = 12, so your birthday is on the 12th.

Card 0			
1	3	5	7
9	11	13	15
17	19	21	23
25	27	29	31

Card 1			
2	3	6	7
10	11	14	15
18	19	22	23
26	27	30	31

Card 2			
4	5	6	7
12	13	14	15
20	21	22	23
28	29	30	31

Card 3			
8	9	10	11
12	13	14	15
24	25	26	27
28	29	30	31

Card 4			
16	17	18	19
20	21	22	23
24	25	26	27
28	29	30	31

The reason this trick works is given in *The Secret Behind the Magic Birthday Predictor Cards* (p. 30). If you would like to figure it out for yourself, first look for a pattern in the numbers appearing on each card.

21

Number Theory

Divisibility Rules

Karl Friedrich Gauss, one of the greatest mathematicians of all time, was one of the few mathematicians who was also a calculating prodigy. (You will learn more about Gauss later on page 137.) Of the four arithmetic operations, Gauss was least fond of division. In his treatise, *Disquisitiones Arithmeticae*, published in 1801 when he was 24 years old, Gauss presented a unification of many disparate results of number theory in a new formulation called *modular arithmetic*. One facet of this new approach to arithmetic was to change division into multiplication where possible and to reduce the amount of division required in computation. In standard arithmetic, we say that two integers are equivalent if and only if they are equal. In arithmetic *modulo n*, we say that two integers are equivalent if and only if they have the same remainder when divided by *n*. More formally, we define *equivalence modulo n* as follows:

Definition: Two integers x and y are said to be *equivalent modulo n* if $x - y$ is divisible by n.

We write $x \equiv y \pmod{n}$ when we mean $x - y$ is divisible by n.

In a nutshell, equivalence modulo n partitions the integers into sets or classes according to the remainder yielded upon division by n. All integers which yield the same remainder are deemed to be equivalent. For example, equivalence modulo 2 partitions the integers into two sets, the even and the odd numbers.

Set of Integers, $\mathbb{Z} = \{0, \pm2, \pm4, \ldots\} \cup \{\pm1, \pm3, \pm5, \ldots\}$

Equivalence modulo 3 partitions the integers into three sets, the multiples of 3, the integers of the form $3n + 1$, and integers of the form $3n + 2$.

Set of Integers, $\mathbb{Z} = \{3n \mid n \in \mathbb{Z}\} \cup \{3n + 1 \mid n \in \mathbb{Z}\} \cup \{3n + 2 \mid n \in \mathbb{Z}\}$

We observe that numbers in the same set are equivalent modulo 3; for example,

$$3 \equiv 45 \pmod{3} \quad \text{and} \quad 5 \equiv 32 \pmod{3}$$

This section will use modular arithmetic and "mod" notation. If you have just learned about it, you might enjoy seeing how they can be used.

Number Theory

Everyone is familiar with rules for divisibility by numbers such as 2, 3, 4, and 5. Divisibility rules for other numbers such as 11 are easy to remember as well. The proofs that these rules actually work are quite easily grasped using some fundamental notions of modular arithmetic. In this section, we'll prove that the usual divisibility rules work, and derive some lesser-known ones. Let's start with the simplest case.

Divisibility by 2

A number is divisible by two if and only if the last digit is even.

This may seem too easy, but it will allow us to develop some tools that we will use later in greater generality. When you see a number (such as 471), you should of course keep in mind that the positions of the digit correspond to powers of 10:

$$471 = 4 \times 10^2 + 7 \times 10^1 + 1$$

If we want to prove something about a general n-digit number M, we need variables to represent its digits. If the digits of M are (from left to right) $a_{n-1}, a_{n-2}, \ldots, a_1, a_0$ then we will write: $M = (a_{n-1}a_{n-2} \cdots a_1 a_0)_{10}$

The subscripted "10" is there to remind us that this shouldn't be read as a product, and that this really means

$$M = a_{n-1} \times 10^{n-1} + a_{n-2} \times 10^{n-2} + \ldots + a_1 \times 10^1 + a_0$$

(We might later want to get fancy and replace 10 by some other number such as 2 or 3. This is known in the lingo as *working in another base*.)

To check whether M is divisible by 2, we need only check that $M \equiv 0 \pmod 2$. Since $10 \equiv 0 \pmod 2$,

$$\begin{aligned} M &\equiv a_{n-1} \times 10^{n-1} + a_{n-2} \times 10^{n-2} + \ldots + a_1 \times 10^1 + a_0 \\ &\equiv a_{n-1} \times 0^{n-1} + a_{n-2} \times 0^{n-2} + \ldots + a_1 \times 0^1 + a_0 \\ &\equiv a_0 \pmod 2 \end{aligned}$$

Can you prove the usual divisibility rules for 5 and 10 using this method?

Thus M is divisible by 2 if and only if its last digit is divisible by 2!

Number Theory

Divisibility by 9

A number is divisible by 9 if and only if the sum of its digits is divisible by 9. (This method can easily be repeated. For example, to test if 1234567890 is divisible by 9, we add up its digits to get 45; we then add up the digits of 45 to get 9, which is certainly divisible by 9.) The well-known method for checking if sums are correct — known as "casting out nines" — relies on this sort of argument as well.

Proceeding as above, we observe that if M is a general n-digit number, we write

$$M = a_{n-1} \times 10^{n-1} + a_{n-2} \times 10^{n-2} + \ldots + a_1 \times 10^1 + a_0.$$

This time, we use the fact that $10 \equiv 1 \pmod 9$, and so $10^k \equiv 1^k \pmod 9$ for all k.

$$
\begin{aligned}
M &\equiv a_{n-1} \times 10^{n-1} + a_{n-2} \times 10^{n-2} + \ldots + a_1 \times 10^1 + a_0 \\
&\equiv a_{n-1} \times 1^{n-1} + a_{n-2} \times 1^{n-2} + \ldots + a_1 \times 1^1 + a_0 \\
&\equiv a_{n-1} + a_{n-2} + \ldots + a_1 + a_0 \pmod 9
\end{aligned}
$$

This proof shows us that we can get even more information than the basic divisibility rule: adding up the digits tells us the value of M modulo 9.

For example,
$$
\begin{aligned}
123454321 &\equiv 1 + 2 + 3 + 4 + 5 + 4 + 3 + 2 + 1 \\
&\equiv 25 \\
&\equiv 2 + 5 \\
&\equiv 7 \pmod 9
\end{aligned}
$$

Try to prove the following divisibility rules:

Divisibility by 3

A number is divisible by 3 if and only if the sum of its digits is divisible by 3. (More generally, you can prove that a number is congruent to the sum of its digits modulo 3.)

Number Theory

Divisibility by 4

> A number is divisible by 4 if and only if the last two digits (taken as a two-digit number) are divisible by 4. (Can you make up an analogous rule for divisibility by 8?)

The above test for divisibility by 4 is equivalent to the following test:

> If the last two digits of a number are ab, then the number is divisible by 4 if and only if one of the following conditions holds:
>   i) b is divisible by 4 and a is even
>   ii) b is even but not divisible by 4, and a is odd

Divisibility by 11

> A number, $M = (a_{n-1}a_{n-2} \cdots a_1 a_0)_{10}$ is divisible by 11 if and only if
>
> $a_0 + a_2 + a_4 + \ldots = a_1 + a_3 + a_5 + \ldots$ (Hint: $10 \equiv -1 \pmod{11}$)

Can you think of a simple method for finding the remainder when you divide $(a_{n-1}a_{n-2} \cdots a_1 a_0)_{10}$ by 11? (Simply dividing by 11 and seeing what you get is not allowed!)

Composite Numbers in General

To test for divisibility by 6, you need only check for divisibility by 2 and by 3. To check for divisibility by 12, you need only check for divisibility by 3 and 4. (But it doesn't suffice to check only for divisibility by 2 and 6, although 2 times 6 is also 12. Why not?) Using this method, you can easily check for divisibility by many new numbers, such as 12, 15, 36, and 60.

Number Theory

The Factor Theorem

EXPERTS ONLY

If you are a seasoned veteran, you will be familiar with the following powerful theorem.

The Factor Theorem

> Let $g(x)$ denote any polynomial in x. Then $(x\text{-}a)$ is a factor of $g(x) \Leftrightarrow g(a) = 0$.

The symbol $\Leftrightarrow$ means "if and only if".

The Factor Theorem relates the zeros of any polynomial to its linear factors. To determine whether $(x - 3)$ is a factor of the polynomial $g(x) = x^3 - 8x^2 + 22x - 18$, we evaluate $g(3)$. Substituting 3 for x yields $3^3 - 8(3)^2 + 22(3) - 18 = 3$. Since $g(3) \neq 0$, $(x - 3)$ is not a factor of $g(x)$.

Can you see how the factor theorem is conceptually related to the divisibility rules for 9 and 11? (Can you see that a polynomial is divisible by $(x\text{-}1)$ if and only if the sum of its co-efficients is 0? For example, $x^{73} - 3x^2 + 2$ is divisible by $(x\text{-}1)$, but $x^{94} + 5x + 3$ isn't. This has the same flavor as our test for divisibility by 9. Can you think of a similar rule to test for the divisibility of a polynomial by $(x + 1)$, that is essentially the same as our divisibility rule for 11?)

The Factor Theorem is a special case of the following more general theorem:

The Remainder Theorem

> Let $g(x)$ denote any polynomial in x. Then the remainder when $g(x)$ is divided by $(x\text{-}a)$ is $g(a)$.

We can prove the remainder theorem by observing that any polynomial, $g(x)$, can be divided by $(x - a)$ to yield a quotient, $f(x)$, and a constant term, say r. Then
$$g(x) = (x - a)f(x) + r$$
Substituting a for x into this equation, we find $r = g(a)$, so $g(x) = (x - a)f(x) + g(a)$
Therefore, on division of $g(x)$ by $(x\text{-}a)$ we obtain quotient, $f(x)$ and remainder $g(a)$.
Do you see how to prove the Factor Theorem using the Remainder Theorem?

If you reflect for a moment, you might see how the Remainder Theorem implies

$$(a_{n-1}a_{n-2} \ldots a_1a_0)_{10} \equiv a_{n-1} + a_{n-2} + \ldots + a_1 + a_0 \,(\text{mod}\, 9).$$

Number Theory

More Complicated Divisibility Tests

Less well-known is the following rule for divisibility by 7.

Divisibility by 7

$(a_{n-1}a_{n-2} \cdots a_1 a_0)_{10}$ divisible by 7 $\Leftrightarrow$ $(a_{n-1}a_{n-2} \cdots a_2 a_1)_{10} + 5a_0$ is divisible by 7.

For example, to check if 12345 is divisible by 7, we can use this rule repeatedly.

12345 is divisible by 7 $\Leftrightarrow$ 1234 + 5 × 5 is divisible by 7
$\Leftrightarrow$ 1234 + 25 is divisible by 7
$\Leftrightarrow$ 1259 is divisible by 7
$\Leftrightarrow$ 125 + 5 × 9 is divisible by 7
$\Leftrightarrow$ 125 + 45 is divisible by 7
$\Leftrightarrow$ 170 is divisible by 7
$\Leftrightarrow$ 17 + 5 × 0 = 17 is divisible by 7 , which it isn't.

As you can see, this rule is just a little better than dividing by 7. But with a few other ad hoc rules, it can be made more efficient. For example, let's check if 1234567 is divisible by 7.

❶ First, we throw out the final 7, which won't affect divisibility by 7. (Why?) This leaves us with 1234560.
❷ Next, we can divide by 10, as that again won't affect divisibility by 7. (Why?) We're left with 123456.
❸ We use the rule to transform 123456 to 12345 + 5 × 6 = 12375.
❹ Ditch the 7 to get 12305.
❺ Use the rule again to get 1230 + 5 × 5 = 1255.
❻ And again: 125 + 5 × 5 = 150.
❼ We're done: we can divide by 10 to get 15, which isn't divisible by 7.

Therefore, 1234567 is not divisible by 7. With a little practice, you will be able to do this test quite quickly.

Now let's prove that this rule works. Let the number in question be M. Instead of having one variable for each digit, it will be easier in this case to have one variable for the units digit, and one variable for the rest of the number. Say $M = 10 x + y$, where x is a non-negative integer, and y is an integer between 0 and 9 inclusive.

Number Theory

We wish to prove the following complicated-looking proposition:

Proposition

> $10x + y$ is divisible by 7 $\Leftrightarrow$ $x + 5y$ is divisible by 7.

Proof.

We can multiply a number by 5 without changing its divisibility by 7. Thus:

$$10x + y \text{ is divisible by } 7 \Leftrightarrow 5(10x+y) \text{ is divisible by } 7$$
$$\Leftrightarrow 7\,(7x) + (x+5y) \text{ is divisible by } 7$$

But we can throw out multiples of 7, so

$$7\,(7x) + (x+5y) \text{ is divisible by } 7 \Leftrightarrow x+5y \text{ is divisible by } 7$$

(In fact, this proof can be very slightly changed to show that we've actually found a divisibility rule for 49. Can you see how?)

Of course, this entire proof can be rephrased in the language of modular arithmetic. For example, the first line would read:

$$10\,x + y \equiv 0 \,(\text{mod } 7) \Leftrightarrow 5(10x+y) \equiv 0 \,(\text{mod } 7)\,.$$

Having seen how the divisibility rule for 7 works, you might be able to prove a similar rule for 13:

Divisibility by 13

$$(a_{n-1}a_{n-2} \cdots a_1 a_0)_{10} \text{ divisible by } 13 \Leftrightarrow (a_{n-1}a_{n-2} \cdots a_1)_{10} + 4\,a_0 \text{ divisible by } 13$$

If you are feeling *really* ambitious, you can find a similar rule for divisibility by 19 (in which the "4" in the 13-rule is replaced by a "2") or for 23, 29, 59, 79, or 89!

Number Theory

Food for Thought

❶. Using the above and similar techniques, you should be able to devise divisibility rules for every number less than 30 except for 17 and 27. (Can you think of a reasonable test for either of these "missing numbers"? Probably not, but it's worth trying!)

❷. If humans had twelve fingers instead of ten, we might be counting in base 12 today. Can you think of simple divisibility tests for 2, 3, and 4 in a base 12 world? How about 11 and 13?

❸. Here's a divisibility test for 7, 11, and 13 — all at the same time! It works well on very large numbers. If you're given a huge number with lots of digits (e.g. 342,032,731,131), the commas naturally divide it into chunks of size three (or less). Just take the first chunk, subtract the second, add the third, and so on. The new number you get should be a lot more reasonable in size. Your original number is divisible by 7 if and only if the resulting number is divisible by 7. (The same trick works for 11 and 13.) In this example, 342 -32 +731 -131 = 910. This is divisible by 7 and 13, but not 11, so 342,032,731,131 is divisible by 7 and 13, but not 11. Why does this work?

❹. (This problem is not really related to this section, but it involves divisibility and digits.) Find a positive integer whose first digit is 1 and which has the property that if this digit is transferred to the end of the number, the number is tripled. Then find some more.

If you're interested in learning more about number theory, flip forward to First Steps in Number Theory (p. 120).

Number Theory

The Secret Behind the Magic Birthday Predictor Cards (p. 21)

The riddle begins to untangle when we write the numbers in base 2 (i. e. in binary form) .

•The numbers on card 0 are those numbers that, when written in base 2, have a "1" in the *units* place.
•The numbers on card 1 are those numbers that have a "1" in the *two's* place.
•The numbers on card 2 are those numbers that have a "1" in the *four's* place.

•In general, the numbers on card k have a "1" in the 2^k place. (That's why the cards are numbered 0 through 4, and not 1 through 5.)

Of course, you may renumber them to make the secret more obscure if you wish; it doesn't change the trick at all. You can also scramble the numbers on each card and then choose the smallest number rather than the one in the upper left corner.

For example, 25 in base 2 is 11001. There are 1's in the 2^0, 2^3, and 2^4 places. So 25 appears on cards 0, 3, and 4.

> **Something to think about.**

You can vary the game by having a person think of a number between 1 and 31. Then ask that person to identify the cards on which the secret number appears. Proceeding as above, you can determine the secret number. How many cards would be needed to play this game if the person can choose any number up to 100?[1] What numbers would appear on these other cards?

Card 0

(1)	3	5	7
9	11	13	15
17	19	21	23
25	27	29	31

Card 1

2	3	6	7
10	11	14	15
18	19	22	23
26	27	30	31

Card 2

4	5	6	7
12	13	14	15
20	21	22	23
28	29	30	31

Card 3

(8)	9	10	11
12	13	14	15
24	25	26	27
28	29	30	31

Card 4

(16)	17	18	19
20	21	22	23
24	25	26	27
28	29	30	31

---

[1] The number of cards required is the number of bits needed to encode a number between 1 and 100. So this simple trick is just a couple of steps away from concepts used in the way computers encode information!

Number Theory

Magic Squares

A magic square of order n is a square array of the natural numbers from 1 to n^2, such that the sum of the numbers in each row, column, and diagonal is the same. The following array is a magic square of order 4, with a constant sum of 34.

1	14	15	4	→	34
8	11	10	5	→	34
12	7	6	9	→	34
13	2	3	16	→	34

	34	34	34	34		34

This constant sum is called the *magic sum* of the magic square. These mathematical objects were sacred to the ancient Chinese, and were thought to have mystical powers. Magic squares of order n exist for all positive integers n except $n = 2$. (It isn't hard to see why there are no magic squares of order 2.)

In general, there are many magic squares of any given order. In the next section, *Two Theorems About Magic Squares,* we will look at a little theory behind them. Meanwhile, here is a method to make magic squares of all possible sizes.

Constructing Magic Squares of Order n (n odd)

We'll construct a magic square of order 5. All other odd squares follow the same pattern.

❶ Place two other 5 x 5 arrays around the central 5 x 5 array, one directly above and one to the right. Start off by placing a 1 in the middle square in the top row of the central 5 x 5 array.

❷ Then place the integers in order from 2 to 25, moving diagonally *up and to the right* at each step. If you run off the central magic square onto an abutting array, go to the same position in the central magic square and record the number.

❸ Every so often (at the multiples of n) you'll be prevented from continuing. In that case, go instead to the square immediately below where you were. (Notice where the 6 was placed in this sample magic square.)

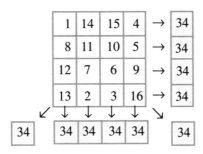

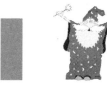

Number Theory

Constructing Magic Squares of Order n (n = 4m+2)

As an example of this case, we'll construct a magic square of order 6 (so $m = 1$). Partition the magic square into four smaller squares A, B, C, D as shown.

A, B, C, and D are each $(2m + 1)$ x $(2m + 1)$ squares. Fill them with identical magic squares of order $2m+1$. (We learned how to do this in the previous case.) Proceeding in this way, we obtain the following 3 x 3 magic square.

8	1	6
3	5	7
4	9	2

When this 3 x 3 magic square is substituted for A, B C and D, our 6 x 6 magic square "under construction" will look like this:

8	1	6	8	1	6
3	5	7	3	5	7
4	9	2	4	9	2
8	1	6	8	1	6
3	5	7	3	5	7
4	9	2	4	9	2

Now, add $(2m+1)^2$ (in this case 3^2 or 9) to each of the elements in C, $2(2m+1)^2$ (in this case 18) to each of the elements of B, and $3(2m+1)^2$ (in this case 27) to each of the elements of D. At this point, the magic square contains the integers from 1 to n^2, and the column sums are all equal, but it isn't really magic yet. Our 6 x 6 magic square under construction becomes:

8	1	6	26	19	24
3	5	7	21	23	25
4	9	2	22	27	20
35	28	33	17	10	15
30	32	34	12	14	16
31	36	29	13	18	11

The final step may seem arbitrary, but if you work out the row and diagonal sums in the square below, you'll see that this is just what the doctor ordered to make the square magical.

In subsquare A (the upper-left one), circle the first *m* numbers in every row but the middle one. (In the example, *m* = 1, so there is only one number) In the middle row, again circle *m* consecutive numbers. but start from the second number. In the example given, you would circle the numbers 8, 4, and 5. Then, in subsquare D, do the same thing (so in the example, the numbers 35, 31, and 32 would be circled). Finally, swap the circled numbers in square A with the corresponding numbers in square D. Our 6 X 6 example becomes:

⑧	1	6	26	19	24
3	⑤	7	21	23	25
④	9	2	22	27	20
㉟	28	33	17	10	15
30	㉜	34	12	14	16
㉛	36	29	13	18	11

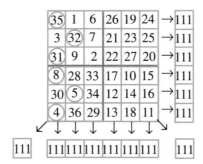

Our square is now magic!

Number Theory

Constructing Magic Squares of Order n (n = 4m)

First of all, write the numbers from 1 to n^2 in the square, filling in the rows from left to right, beginning with the top row and working down. (In the following example we construct a magic square of order $n = 8$, i.e. $m = 2$.)

1	2	3	4	5	6	7	8
9	10	11	12	13	14	15	16
17	18	19	20	21	22	23	24
25	26	27	28	29	30	31	32
33	34	35	36	37	38	39	40
41	42	43	44	45	46	47	48
49	50	51	52	53	54	55	56
57	58	59	60	61	62	63	64

1	2	3	4	5	6	7	8
9	10	11	12	13	14	15	16
17	18	19	20	21	22	23	24
25	26	27	28	29	30	31	32
33	34	35	36	37	38	39	40
41	42	43	44	45	46	47	48
49	50	51	52	53	54	55	56
57	58	59	60	61	62	63	64

Divide the magic squares into m^2 subsquares of order 4. In each 4 x 4 subsquare, circle the diagonal entries as shown in the figure above right.

Finally, in the large square, replace all of the circled numbers by switching each one with the "opposite" entry in the magic square (with respect to the center of the square). That is, interchange all the circled numbers with their images under a rotation of the magic square 180° about its center. For example, the entry in the third row and second column (i.e. 18) would be switched with the entry in the third-last row and second-last column (i.e. 47). (If this is too confusing, you can replace any circled number x with the number $n^2 + 1 - x$.) Our completed magic square is shown in the figure on the right.

64	2	3	61	60	6	7	57
9	55	54	12	13	51	50	16
17	47	46	20	21	43	42	24
40	26	27	37	36	30	31	33
32	34	35	29	28	38	39	25
41	23	22	44	45	19	18	48
49	15	14	52	53	11	10	56
8	58	59	5	4	62	63	1

There are many places to read more about magic squares. *Mathematical Recreations and Essays* by Coxeter and Rouse Ball is an excellent source. See the *Annotated References* for bibliographic information.

Number Theory

Two Theorems about Magic Squares

In *Magic Squares,* we saw how to construct magic squares of any order greater than 2. In this section, we will prove a couple of theorems about them.

Theorem 1

> In a magic square of order n, the magic sum is $\dfrac{n(n^2+1)}{2}$.

In *Magic Squares*, we had three different methods for creating magic squares. You can check that in each of the examples we created (n = 4, 5, 6, and 8) that the theorem holds. (This makes us happy, because otherwise the theorem wouldn't be very good!)

Proof.

Let s denote the magic sum for a magic square of order n. Since this is a magic square, the sum of the elements in each of its n rows is s. Therefore the sum of the elements in all n rows is ns.

But, the elements in the magic square are the numbers 1, 2, 3, ..., n^2, so the sum of the elements in the magic square is the sum of the first n^2 natural numbers; that is,

$$\frac{n^2(n^2+1)}{2}.$$

This formula will be proved in *Sums of kth Powers* (p. 135).

Combining the two preceding statements, we have $ns = \dfrac{n^2(n^2+1)}{2}$ from which we deduce that $s = \dfrac{n(n^2+1)}{2}$.

So our first theorem tells us, for example, that the magic sum of a 10 × 10 magic square is 505, before we even look at a magic square of order 10.

Number Theory

Our second theorem says that there is *essentially* only one magic square of order 3. Before we discuss it, we'll have to elaborate on what we mean by "essentially". For example, at first glance, the following magic squares all look different.

I

8	1	6
3	5	7
4	9	2

II

6	1	8
7	5	3
2	9	4

III

4	3	8
9	5	1
2	7	6

But square II is obtained from square I by switching the first and third columns — in effect, by reflecting the square in the middle column. Square III is obtained from square I by rotating the square clockwise 90 degrees. (It can also be obtained from square II by reflection in the diagonal passing from the top-right of the square to the bottom-left.)

We say that these squares are identical *up to rotations and reflections*. (This is one instance of the mathematical concept of symmetry.)

Theorem 2

> There is only magic square of order 3 (up to symmetry).

One way of proving this is to try all possible ways of placing the integers 1 through 9 into a 3 x 3 matrix. There are $9! = 362,880$ possibilities to try, so this would take some time. With a bit of cleverness, we can avoid this hassle.

Proof.

Step 1 The magic sum must be 15. (This is just Theorem 1 for the special case $n = 3$.)

Step 2 The number in the middle square must be 5.

The proof of this is very similar to the proof of Theorem 1. Fill up the magic square with 9 unknowns $a, b, c, d, e, f, g, h, i$ as follows:

a	b	c
d	e	f
g	h	i

Now, for example, $a+e+i = 15$, because the sum of the elements on the main diagonal must equal the magic sum. Therefore,

$$3e = (a+e+i) + (b+e+h) + (c+e+g) - (a+b+c) - (g+h+i)$$
$$= \quad 15 \quad + \quad 15 \quad + \quad 15 \quad - \quad 15 \quad - \quad 15$$
$$= \quad 15$$

So $e = 5$.

From here, there are several directions to go. If you are feeling intrepid, you may want to explore on your own before continuing.

Step 3 The numbers 9 and 8 can't be in the same row, column, or diagonal, because their sum is already greater than 15, the magic sum. Similarly, the 9 and 7 can't be in the same row, column, or diagonal, and neither can the 8 and 7. So where can these three "big" numbers fit? A little thought shows that they must lie in one of the following configurations (where the big numbers are represented by a "B").

B		
	5	B
	B	

		B
B	5	
	B	

	B	
B	5	
		B

	B	
	5	B
B		

Step 4 Now, the 4 and 6 must lie on opposite sides of the 5; 4 -5- 6 must be either a row, column or diagonal passing through the center. But, looking at where the big numbers have to be placed (see step 3), we see that the 4-5-6 must be a diagonal. As we're considering magic squares to be the same up to rotations, we can assume that the 4 is in the top right corner. So our magic square looks like:

		4
	5	
6		

Step 5 Where can the 9 be now? The 9 can't be in the same row or column as the 6, as their sum is already 15, and the addition of the third term in that row or column would knock the sum over 15. So the 9 must be in one of the circled positions in the following diagram.

	⭕	4
	5	⭕
6		

But we're counting magic squares to be the same up to reflection, so we may as well assume the 9 is at the middle position of the first row. Our magic square now looks like:

	9	4
	5	
6		

Step 6 From here, using the fact that magic sum is 15, it's easy to fill out the remaining elements. So the only magic square, up to rotations and reflections, is:

2	9	4
7	5	3
6	1	8

We observe that this is the same as the 3 x 3 magic square in *Magic Squares* (p. 32), rotated by 180° about the center.

Number Theory

❶. If we take away the "symmetry" condition, how many magic squares of order 3 are there? (Answer: 8)

❷. In how many ways can 15 be expressed as the sum of three distinct positive integers between 1 and 9? Do all such triplets appear as rows, columns, or diagonals of the 3 x 3 magic square? (This fact can be used to provide another proof of Theorem 2.)

❸. **The Magic Fifteen Game**[1]
Nine cards, the ace through nine of spades, are removed from the deck and placed face up on a table. Two players alternate in picking up cards. The first player to have three distinct numbers summing to fifteen wins. (Some games will end in a draw.) Can you see a trick that will enable you to play really well? (Hint #1: It is identical to another well-known game mentioned elsewhere in this book. Hint #2: Why has this problem been included in this particular section?)

---

[1]In Berlekamp, Conway, and Guy's *Winning Ways*, **The Magic Fifteen Game** is attributed to E. Pericoloso Sporgersi.

Do What You Enjoy!

J.P. Grossman (Canada)
Born March 23, 1973

J.P. Grossman's playful approach to life was evident from an early age. In kindergarten, he developed an enduring interest in taking things apart: clocks, calculators, watches, radios — even computers and a mechanical record player. His ingenuity didn't always meet with popular approval, however; he was once sent to the principal's office for working too far ahead of his class.

"Taking things apart" to understand them typifies J. P.'s method of learning. In Grade 9, he and a friend read an article in *Discover* magazine about fractals, at the start of the fractal craze. They were curious to find out how fractals worked. The first hurdle was to understand this mysterious "i" that appeared in the equations as the alleged square root of -1. Rather than ask for advice and risk being told that their questions were "too hard", they played around with the concept of "i" and figured out how it worked. They then wrote computer programs to generate fractals on whatever machines they could find, starting with a Vic 20 and working their way up, beginning with BASIC and learning other languages as they needed them.

In Grade 10, J. P.'s competition career took off, thanks to his intuitive approach to mathematical challenges combined with his ability to pick problems apart. In Grade 11, he won the *Canadian Mathematical Olympiad* for the first of three consecutive years, easily qualifying for the Canadian Team to the *International Mathematical Olympiad*. During his three years competing at the IMO, he won two Silver Medals and a Gold Medal. Moreover, in his second-last year of high school he placed first in the *USA Mathematical Olympiad*, the last year Canadians were officially allowed to compete. While studying mathematics, physics, and electrical engineering at the University of Toronto, J.P. took the North American *Putnam Mathematical Competition* three times, twice achieving the highest honor of Putnam Fellow. He has one more year of eligibility before he graduates from Toronto.

Personal Profile

Along with his talents in mathematics, J. P. pursues many other interests, including chess, debating, competitive soccer, and skiing. A jazz enthusiast, he is also an accomplished tenor saxophone player.

After graduation, J. P. plans to work in VLSI (Very Large Scale Integration) chip design. For J. P., chip design is problem solving of a different sort, requiring (like mathematics) tinkering, fiddling around, and taking things apart.

What J. P. loves most about mathematics are the ingenious leaps required, "putting together two facts that seem unrelated." He finds that he has become experienced in "making connections between different things just by recognizing certain patterns," both within and outside the field of mathematics. One of his favorite examples of this principle is *The Magic Birthday Predictor Cards* (p. 21). When he saw this trick at a young age and realized how it worked, he had a sudden insight into binary numbers.

J. P.'s curiosity is typical of the young mathematicians profiled in this book. Also typical is the vigor with which he pursues his interests. Rather than seeing his undiscriminating intellectual acquisitiveness as a distraction that holds him back and wastes his time, he finds it constantly driving him forward. His philosophy of life? "Do what you enjoy!"

COMBINATORICS

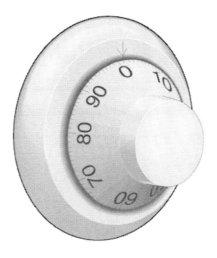

Music is the pleasure the human soul experiences from counting without being aware that it is counting.

— Gottfried Leibniz

Combinatorics

A Mathematical Card Trick

This card trick is incredible when you first see it. But be prepared for a let-down when you figure out how it works!

The Trick

I ask you to shuffle a deck of cards thoroughly. Then I ask for them back (face down). Carefully examining the backs of the cards, I separate them into two piles. I then claim that, through the power of magic, I've made sure that the number of black cards in the first pile is the number of red cards in the second pile!

How did I do it?!

The answer is in ***How the Mathematical Card Trick Works*** *(p. 51). But don't flip there until you've spent some time thinking about it!*

Combinatorics

Counting the Faces of Hypercubes

Combinatorics, essentially the art and science of counting, has recently become one of the most active fields in mathematics. Some of you may know it as "finite mathematics", a subject often introduced at the end of high school. It exists at the crossroads of mathematics and computer science; many combinatorial results are used to determine and evaluate algorithms.

More important (to me, at least) is its aesthetic appeal. The field of combinatorics contains many surprising results and elegant proofs. Best of all, it doesn't require years of preparatory study; elementary results are pretty much accessible to anyone. Some of the aspects we will examine are Pascal's Triangle, the theory of games, and other diversions.

To begin with, let's count something that you wouldn't often think of counting. How many faces does a cube have? Thinking of a die, you would probably answer, "Six." Not too exciting.

So let's make our definition of *face* more flexible. The faces we commonly think about when we consider a cube are two-dimensional. Let's include *faces* of all dimensions. For example, there are 12 edges on a cube. (Think about it!) Let's call these one-dimensional faces. There are also 8 zero-dimensional faces: the vertices of the cube. The cube is a three-dimensional object, so it doesn't have any faces of dimension four or higher. But (watch out, this is weird) it does have one three dimensional face — the entire cube!

So the total number of faces on a cube is:

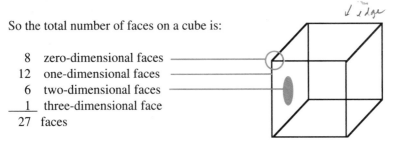

✓ edge

8	zero-dimensional faces
12	one-dimensional faces
6	two-dimensional faces
1	three-dimensional face
27	faces

This is still not too exciting, so let's generalize some more definitions. We've generalized "faces" — now let's generalize "cube"! What we usually think of as a cube, we'll now call a "three-dimensional cube". The reasonable candidate for a two-dimensional cube is a square. Continuing the pattern, we take a line segment to be a one-dimensional cube, and a single point as a zero-dimensional cube.

Combinatorics

Using the definition, we can even define a four-dimensional cube (a "hypercube"), or even higher-dimensional cubes! But we'll talk about such things later.

Now let's count faces on our smaller-dimensional cubes. Make sure I'm not pulling the wool over your eyes by thinking through all of these numbers yourself.

Zero-dimensional cube	One-dimensional cube	Two-dimensional cube
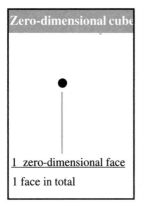	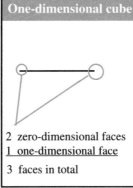	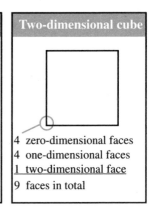
1 zero-dimensional face	2 zero-dimensional faces	4 zero-dimensional faces
	1 one-dimensional face	4 one-dimensional faces
		1 two-dimensional face
1 face in total	3 faces in total	9 faces in total

So now we have a table:

dimension of cube	0	1	2	3	4	5
number of faces	1	3	9	27	?	?

The pattern is clear! So now we can say, with reasonable confidence, that the *hypercube* (four-dimensional cube) has 81 faces! But first we must define what we mean by a *hypercube of dimension n*. Since a cube is a 3-D figure traced out by a square as it moves in a third dimension, we can imagine a 4-D figure traced out by a cube as it moves in a fourth dimension "perpendicular" to the other dimensions. Such a cube we may refer to as a *hypercube of dimension 4*. In general, we can define a hypercube of any dimension n in terms of a hypercube of dimension $n - 1$.

Definition: A *hypercube* of dimension n is a geometric figure traced out when a cube of dimension $n - 1$ is moved in the direction of the nth dimension.

Another possibiliy would be to define the n-dimensional hypercube to be the set

$$\left\{ (x_1, x_2, ..., x_n) \mid 0 \le x_1, x_2, ..., x_n \le 1 \right\}$$

You can see how this would give what we want in dimensions 1, 2, and 3. Use the definition you like best.

Combinatorics

Now we have a hypothesis that we want to prove.

> **Hypothesis:** The n-dimensional cube has 3^n faces.

Proof.

This proof uses the method of *mathematical induction*. (If you've never heard of this, don't run away in fright; you will be able to figure out what's going on.)

When $n = 0$, the result is true; that is, the zero-dimensional cube has 3^0 faces.

We'll now show that if the result holds for an arbitrary n, then it holds for $n+1$. (We've shown it for $n = 0$, and now, using this fact repeatedly, we can show it for $n = 1$, $n = 2$, $n = 3$, ... as far as we'd like to go. This, in essence, is the idea behind mathematical induction.)

As an example, consider the cube (of dimension 3) traced out by moving the colored square (a cube of dimension 2) out of the plane of the page as shown in the diagram. We observe that the colored square has 4 faces of dimension 0 (i.e. 4 vertices). As the colored square moves out of the plane of the page, vertex A moves to position A', creating another face of dimension 0, plus a face AA' of dimension 1. That is, each face of dimension 0 in the colored square contributes 3 faces to the cube.

Similarly, the line segment AB, contributes to the cube, two line segments AB and A'B' (faces of dimension 1) plus face AA'B'B of dimension 2.

In general, as a hypercube of dimension n traces a hypercube of dimension $n + 1$, each face of dimension k (where $k \leq n$) in the hypercube of dimension n contributes 3 faces to the hypercube of dimension $n + 1$: itself, its new image and the new face joining its initial and final positions. All faces are created in this way.

Since every face of the hypercube of dimension n contributes 3 faces to the hypercube of dimension $n + 1$ as it traces that hypercube, and since the hypercube of dimension n was assumed to have 3^n faces, then the hypercube of dimension $n + 1$ has 3×3^n, or 3^{n+1} faces. It follows by induction that the hypothesis is true for all values of n.

Combinatorics

EXPERTS ONLY

We can make the ideas in the foregoing heuristic proof a little more formal, and in the process gain a little more information. For example, we can deduce that the 5-cube has 80 two-dimensional faces.

Let's denote by $F_n(k)$ the number of faces of dimension k possessed by an n-cube. When an n-cube is traced out by moving an $(n-1)$-cube in the new dimension, each face of dimension k contributes 3 faces to the n-cube: itself, its new image (i.e. two faces of dimension k) and the new face of dimension $k+1$ which joins its initial and final positions. This relationship can be expressed in our new notation as follows:

$$F_n(k) = 2F_{n-1}(k) + F_{n-1}(k-1) \quad (*)$$

We can use this relationship to fill in the following table. Observe that the element in row k, column n is obtained by adding the element in row $(k-1)$, column $(n-1)$ to double the element in row k, column $(n-1)$. This is a direct translation of equation $(*)$ into English.

Number of Faces of Each Dimension for n-Cubes; $n = 0,...,5$						
Dimension of the face	0-cube (point)	1-cube (segment)	2-cube (square)	3-cube (cube)	4-cube	5-cube
0	1	2	4	8	16	32
1		1	4	12	32	80
2			1	6	24	80
3				1	8	40
4					1	10
5						1
Total number of faces	1	3	9	27	81	243

Equation $(*)$ above applied to the induction assumption, $\displaystyle\sum_{k=0}^{n} F_n(k) = 3^n$, yields:

$$\sum_{k=0}^{n+1} F_{n+1}(k) = \sum_{k=0}^{n+1} [2F_n(k) + F_n(k-1)]$$

$$= 2\sum_{k=0}^{n+1} F_n(k) + \sum_{k=0}^{n+1} F_n(k-1) \quad \leftarrow \text{Note: } F_n(-1) = 0 \text{ and } F_n(n+1) = 0.$$

$$= 3\sum_{k=0}^{n} F_n(k) \quad = \quad 3^{n+1} \qquad \text{so the hypothesis is proved for all } n.$$

Combinatorics

The best kind of proof is one that gives you a good intuitive idea of what's going on. The problem is, different proofs often work for different people. Which proof did you prefer?

If you find this sort of problem interesting, here are some more problems for you to think about, along a similar vein. These are good problems to ponder when you have nothing better to do, such as when you're walking somewhere, waiting in line, or sitting in the dentist's chair.

Food for Thought

EXPERTS ONLY

❶. a) How many 0-dimensional faces (i. e.vertices) does an n-dimensional cube have? (This isn't too hard.) Remember to prove your answer!
Hint: You could use the definition of an n-cube as the figure in n-dimensional Euclidean space with vertices, the set of points,
$$\{(e_1, e_2, e_3, ..., e_n) \mid e_i = 0 \text{ or } 1\}$$

b) If you're feeling ambitious, and you have some experience with these matters, you can try to prove that the number of k-dimensional faces on an n-dimensional cube is given by:

$$F_n(k) = \binom{n}{k} 2^{n-k} \qquad \text{where } \binom{n}{k} \text{ is defined by } \binom{n}{k} = \frac{n!}{k!(n-k)!}$$

❷. a) A tetrahedron is a triangular pyramid. Let's generalize this idea. What is a two-dimensional tetrahedron? A one-dimensional tetrahedron? A zero-dimensional tetrahedron? In the lingo, a tetrahedron (in any number of dimensions) is called a *simplex*; the plural is *simplices* (which rhymes with *disease*).

b) How many vertices does an n-dimensional simplex have?

c) How many faces (of all dimensions) does an n-dimensional simplex have? (Remember, there are two steps here. First look at small cases and try to guess what the answer should be. Then prove that your guess is in fact correct.)

❸. If you hang a 3-dimensional cube from one of its vertices and slice it horizontally between its highest and lowest vertices, you will get a regular hexagon. Can you see this?

An easier question is, "What happens if you do this with a 2-cube?" A much harder question is: "What happens if you do this with a 4-cube?" (Just as the middle slice of a 3-cube was 2-dimensional, the middle slice of a 4-cube will be 3-dimensional.) The answer is one of the five *platonic solids*.

With these problems, as with others in this book: Experiment! Try small cases! Guess an answer! Play around!

Combinatorics

How the Mathematical Card Trick Works

A Mathematical Card Trick was described on page 44. While pretending to examine the backs of the card, I was simply counting them. I made sure that there were 26 cards in each pile. Why does this work?! You might want to think about this yourself before reading on.

There are several ways of solving this problem. Choose the one that most appeals to you.

Solution 1: (Algebraic)

Let a denote the number of black cards in the first pile, b the number of red cards in the first pile, c the number of black cards in the second pile, and d be the number of red cards in the second pile. We want to show that $a = d$.

There are 26 red cards in the deck, so $b + d = 26$.
There are 26 cards in the first pile, so $a + b = 26$.
Subtracting the second equation from the first, we get: $d - a = 0$.

Solution 2: (Intuitive)

There are 26 cards in each pile. Now move the black cards in the first pile into the second pile. Then move the red cards in the second pile into the first pile. The first pile now contains all the red cards, and the second pile now contains all the black cards. There are still 26 cards in each pile, so the number of cards (black) you moved from the first pile into the second, must be the same as the number of cards (red) you moved from the second to the first.

Combinatorics

If you were intrigued by the mathematical card trick, take a look at the following puzzle.

The Water and Wine Puzzle

Take two glasses of equal capacity. Pour wine into the first glass until it is half full. Pour water into the second glass until it is half full. Take a spoon of wine from the first glass, and put it in the second glass. Then, without worrying about mixing the contents of the second glasses well, take a spoonful of the mixture and put it in the first glass.

Is there more wine in the water glass, or more water in the wine glass?

The answer isn't explicitly given in this book. But if you can't figure it out, read **A Mathematical Card Trick** *(p. 44), and figure out why the situations are really the same.*

Combinatorics

Questions about Soccer Balls

Combinatorics is often about counting something in two different ways. See if you can use this technique to quickly solve the following puzzle:

Twenty-four soccer teams participate in a knock-out tournament. At each stage, the teams pair up and play games, and the losers are eliminated. (If there are an odd number of teams at a stage, one team is chosen by lot to get a "bye" and not play.) At the end of the day, the last team left is declared the victor.

    a) How many games will be played?
    b) What is the answer if there are n teams?

•How many faces are there on a soccer ball?
•How many edges?
•How many vertices? (Vertices are those points where three faces meet.)
• If V is the number of vertices, E is the number of edges, and F is the number of faces, then calculate V-E+F. (Your answer should be 2.)
•Try this same procedure on a cube, a tetrahedron, a triangular prism, a square pyramid, an octahedron, and other polyhedra. What do you notice?

In 1985, a new form of carbon was found, called *Buckministerfullerene* (C_{60}), or *Buckyball* for short. (It is the most stable of a whole family of carbon molecules, now known as *Fullerenes*.) Buckministerfullerene is in the shape of a soccer ball, with the carbon atoms at the vertices and bonds at the edges. Since then, some mathematicians have been studying the shape of the lowly soccer ball, and discovering profound properties which have helped chemists predict how Buckministerfullerene should behave.[1]

DILBERT®   by Scott Adams

---

[1] Robert F. Curl and Richard E. Smalley, "Fullerenes", *Scientific American.* (Vol. 264, No. 4, (Oct. 1991).), 54-63.

Cracking Intractable Problems — That's a Big Buzz

**Catriona Maclean
(United Kingdom)
Born August 15, 1976**

Born and raised in Harrogate, Yorkshire, Katy Maclean attended Harrogate Grammar School, the local state comprehensive school. Phillip Heyes, one of her teachers, introduced her to the excitement of mathematics: "He truly loved his subject, and spent all his time trying to communicate it to his pupils." It was then that Katy started to realize that "there was much more to maths than numbers, and there was some kind of rather beautiful structure behind it all."

Although Katy had always liked mathematics and enjoyed logic puzzles, she came to realize her mathematical ability relatively late in her development. Two years before leaving high school, she did well enough on the *British Mathematical Olympiad* (BMO) to make the national team to the *International Mathematical Olympiad* (IMO), surpassing others who had been marked for success for years. As the dark horse of the team, she won a Silver Medal at the 1993 IMO in Istanbul, Turkey. The next year, she continued to perform spectacularly; she wrote a nearly flawless paper in the *British Mathematical Olympiad*, and went on to win a Gold Medal at the 1994 IMO in Hong Kong with a perfect paper.

It was at this time that Katy discovered the tremendous enjoyment she gains from solving problems: "There are problems that you work at for ages that seem entirely intractable and refuse to respond to anything you hit them with. Then you suddenly dream up some incredibly cute and cunning way of doing them while you're thinking about something completely different. ... Now that's a big buzz."

Since 1994, she has been studying at Emmanuel College, Cambridge, where she is rapidly developing an obsession with arts cinema. True to character, she has also become involved with numerous campus organizations, including the Cambridge University Oxfam group.

Aesthetics is central to Katy's perception of mathematics. She finds herself "inspired by a child-like sense of wonder at the beauty of patterns inherent in the universe. ... I'm not sure whether I think maths discovers beautiful works of nature or creates beautiful works of art. More and more, I tend to think that it does both."

THE FIBONACCI SEQUENCE

*Every new body of discovery is mathematical in form,
because there is no other guidance we can have.*

— Charles Darwin

The Fibonacci Sequence

Elvis Numbers

Elvis the Elf skips up a flight of numbered stairs, starting at step 1 and going up one or two steps with each leap. Along with an illustrious name, Elvis' parents have endowed him with an obsessive-compulsive disorder. Elvis wants to count how many ways he can reach the nth step, which he calls the nth Elvis number, E_n.

Can you figure out the numerical value of E_n? (Hint: it's a well-known sequence.) If you're stuck, try to work out what E_1, E_2, E_3, E_4 and E_5 are, and guess a pattern.

When you've found the answer, or have ended up truly stumped in the effort, flip to *The Solution to Elvis the Elf's Eccentric Exercise* (p. 59).

Fibonacci: The Greatest European Mathematician of the Middle Ages

Leonardo of Pisa (c. 1180-1250) is widely acknowledged to have been the greatest European mathematician of the Middle Ages. He is known to us today as "Fibonacci", or "son of Bonaccio". The son of an Italian merchant, Leonardo studied under a Muslim teacher and made travels to Egypt, Syria, and Greece. As a result, he picked up much of the mathematics that was being disseminated through centers of learning in the medieval Muslim world; in particular, he became a key proponent of Hindu-Arabic numerals, from which were derived the digits 1 through 9.

Courtesy of Columbia University Library

Leonardo da Pisa (Fibonacci) c. 1180 - 1250

Leonardo's classic book, *Liber Abaci* (Book of the Abacus), completed in 1202, discusses numerous arithmetic algorithms, often explained through problems. Certainly the most famous problem in *Liber Abaci* is:

> How many pairs of rabbits will be produced in a year, beginning with a single pair, if in every month each pair bears a new pair which becomes productive from the second month on?

By drawing a tree diagram (see overleaf) showing the number of pairs of rabbits at the end of each month, we discover that the number of pairs of rabbits in the first six months are given by the following sequence, 1, 1, 2, 3, 5, 8, ... This famous sequence, named the *Fibonacci sequence* in honor of Leonardo of Pisa, has the property that each term is the sum of the two previous terms.[1]

In the tree diagram on the following page, we mark the first pair of rabbits with the number 1. Since they do not produce offspring in the first two months, there is only one pair at the end of the second month. On the first day of the third month, they give birth to another pair (labelled 2) so there are two pairs at the end of the third month. The arrows in the tree diagram show the offspring in the successive months up to the end of the sixth month.

---

[1]There is actually a journal, *The Fibonacci Quarterly*, devoted specifically to the Fibonacci numbers.

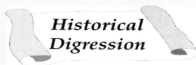
End of Month	The Fibonacci Rabbit Problem	Number of Pairs
1		1
2		1
3		2
4		3
5		5
6		8

↔ newborn pair

Sequences are often defined by an equation relating the *nth term* to the previous terms. In such a case, we say that the sequence is defined *recursively*. The Fibonacci sequence can be defined recursively by the equation,

$$F_n = F_{n-1} + F_{n-2}$$

where F_n denotes the nth term of the sequence. Can you see why the number of pairs of rabbits in the nth month should satisfy this relation?

Try This

If you have a graphing calculator with a sequence mode define the Fibonacci sequence recursively using the equations:

$$u_n = u_{n-1} + v_{n-1} \quad \text{and} \quad v_n = u_{n-1}$$

Then evaluate $\lim_{n \to \infty} \dfrac{u_{n+1}}{u_n}$. Do you recognize this number?

The Fibonacci Sequence

The Solution to Elvis the Elf's Eccentric Exercise

Elvis the Elf's eccentric exercise was posed in *Elvis Numbers* (p. 56). The Fibonacci numbers $F_1, F_2, F_3, \ldots$ are defined as follows.

$$F_1 = F_2 = 1$$
$$F_n = F_{n-1} + F_{n-2} \text{ for } n \geq 3$$

Q. E. D.

The nth Elvis number is actually the nth Fibonacci number! How does Elvis see this? Rather than skipping up the stairs as is his wont, Elvis steps carefully up one at a time. On the nth step, he writes down E_n.

Elvis begins by laboriously counting the number of ways of climbing to the nth step in his green suede shoes:

$E_1 = 1$

$E_2 = 1$

 or

$E_3 = 2$

 or or

$E_4 = 3$

To avoid intense boredom, Elvis thinks of a way to work out E_5 without doing lots of counting. So he thinks, "How can I get to the fifth step? Well, I can get to the third step, and then jump up two steps. There are E_3 ways of doing that. Then again, I might get to the fourth step, and then jump up one step. There are E_4 ways of doing that. But I have to do exactly one of those two things, so the number of ways I can get to step 5 is $E_3 + E_4$. Thus $E_5 = E_4 + E_3$."

Elvis quickly generalizes this procedure to show that $E_n = E_{n-1} + E_{n-2}$ for $n > 2$. Along with $E_1 = 1$ and $E_2 = 1$, this is enough to determine E_n in general. But the Fibonacci numbers are given by the same recipe ($F_1 = 1$ and $F_2 = 1$, and $F_n = F_{n-1} + F_{n-2}$), so these two sequences must be the same! Elvis finishes his proof with the letters Q.E.D., which is short for *Quod Erat Demonstrandum*, a Latin phrase meaning "which was to be proved". It's just a slightly pompous way of showing when a proof is complete.

The Fibonacci Sequence

A Formula for the nth Term of the Fibonacci Sequence

A few pages earlier, we observed that the Fibonacci numbers, F_n, are defined according to the following rules:

$$F_1 = F_2 = 1$$
$$F_n = F_{n-1} + F_{n-2} \text{ for n} \geq 3$$

Then, remarkably, it can be shown that F_n is given by the following equation, often called *Binet's Formula*:

$$F_n = \frac{\left(\frac{1+\sqrt{5}}{2}\right)^n - \left(\frac{1-\sqrt{5}}{2}\right)^n}{\sqrt{5}}$$

This is quite a mouthful!

> **Teachers!**
> Give your students Binet's formula. Invite them to guess the sequence — it's quite a surprise to find nice Fibonacci numbers coming out of that hideous formula! Then they can try to prove it.

We can express F_n more simply using some constants that we'll pull out of our hat:

$$\tau = \frac{1+\sqrt{5}}{2}, \qquad \sigma = \frac{1-\sqrt{5}}{2}$$

τ and σ are Greek letters, called *tau* and *sigma* respectively. *Tau* (which rhymes with "Ow!") is a constant known to the ancient Greeks as the "golden mean". It has a habit of turning up in the oddest places in mathematics; some of them are described in the chapter *Fibonacci & The Golden Mean*. In terms of τ and σ:

$$F_n = \frac{\tau^n - \sigma^n}{\sqrt{5}}$$

There are several other ways of expressing F_n that you might prefer.

Since $\sigma = -\frac{1}{\tau}$ (Check this!), $\qquad F_n = \frac{\tau^n - \left(\frac{-1}{\tau}\right)^n}{\sqrt{5}}$

σ can be expressed in terms of τ in a second way: $\sigma = 1 - \tau$. (Check this too!) Thus

$$F_n = \frac{\tau^n - (1-\tau)^n}{\sqrt{5}}$$

We know from the definition of the Fibonacci sequence (although not from the formula for its general term) that F_n is a positive integer for all n. It follows from the equation above that $F_n = \dfrac{\tau^n}{\sqrt{5}} - \dfrac{\sigma^n}{\sqrt{5}}$ is an integer for all n. Since $\left|\dfrac{\sigma^n}{\sqrt{5}}\right| < \dfrac{1}{2}$ for all n (verify this for yourself), $\dfrac{\tau^n}{\sqrt{5}}$ is closer to F_n than it is to any other integer.

That is, F_n is the integer closest to $\dfrac{\tau^n}{\sqrt{5}}$. Since $\dfrac{\sigma^n}{\sqrt{5}}$ becomes very small as n gets large, $\dfrac{\tau^n}{\sqrt{5}}$ gets very very close to F_n. (Try it on your calculator with $n = 20$.[1])

You can prove that the crazy formula for F_n actually works by employing the same method we used in the *Solution to Elvis the Elf's Eccentric Exercise* (p. 59) — the method of mathematical induction (although induction wasn't actually mentioned by name in that section). Basically, induction is a great way of proving things when you already know (or have guessed) what the answer is.

Using the magical formula, or induction, or a lot of ingenuity, it is possible to prove some remarkable results about Fibonacci numbers, which appear in *Funny Fibonacci Facts* (p. 146).

Food for Thought

How many ways can you spell "ELVIS"?

Start at the E and move either down or to the right at each stage.

```
E  L  V  I  S
L  V  I  S
V  I  S
I  S
S
```

The answer is given on the following page.

---

[1] It is indeed remarkable that for any large n, $\tau^n / \sqrt{5}$ is very very close to an integer. Even more remarkable is that for any large n, τ^n is *also* very very close to an integer — try it on a calculator and see! These oddities, and others, come up in *Some Interesting Properties of the Golden Mean* (p. 151).

Answer to Food for Thought on the Previous Page

There are 16 ways to spell Elvis. Is it a coincidence that the answer turned out to be a power of 2? As a hint, find out how many ways there are to spell "NO" in the following diagram, starting with the "N" in the upper left-hand corner, and moving either down or to the right.

$$N \to O$$
$$\downarrow$$
$$O$$

If you're not convinced (and you're probably not), then count the number of ways of spelling "NEGATIVE":

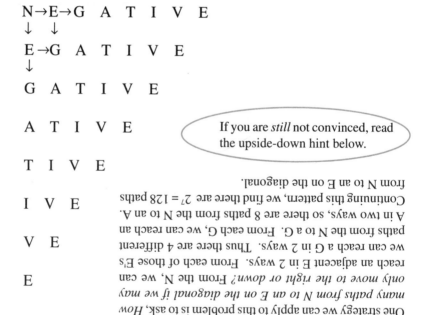

N→E→G A T I V E
↓ ↓

E →G A T I V E
↓

G A T I V E

A T I V E

T I V E

I V E

V E

E

If you are *still* not convinced, read the upside-down hint below.

One strategy we can apply to this problem is to ask, *How many paths from N to an E on the diagonal if we may only move to the right or down?* From the N, we can reach an adjacent E in 2 ways. From each of those E's we can reach a G in 2 ways. Thus there are 4 different paths from the N to a G. From each G, we can reach an A in two ways, so there are 8 paths from the N to an A. Continuing this pattern, we find there are $2^7 = 128$ paths from N to an E on the diagonal.

Here is another question: there are eight E's in the diagonal above. How many paths are there ending at each particular E? Answer: 1-7-21-35-35-21-7-1. How and why does this relate to Pascal's Triangle (discussed in *The Triangles of Pascal, Chu Shih-Chieh, and Sierpinski*, p. 98)?

GAME THEORY

The game isn't over till it's over.

— Yogi Berra

Game Theory

The Theory of Games

Very loosely, *game theory* is the field of mathematics that analyzes strategies in a wide variety of situations, such as the competition for scarce resources. It is very important in economics, and also of interest in political science, psychology and philosophy. One of the 1994 Nobel Prizes in Economics was awarded to John Nash, a mathematician who laid much of the groundwork for game theory in his 1950 Princeton Ph.D. thesis, completed when he was 21. (This wasn't the first Nobel Prize in Economics given for brilliant mathematical work; for another example, see *Game Theory and Politics: Arrow's Theorem*, p. 68.)

One important example of game theory in action is "The Prisoner's Dilemma". There are many variations of this problem, but the basic structure is as follows. Two suspects are taken into custody and separated. The District Attorney is certain that they are guilty of a specific crime, but she does not have enough evidence to convict them at a trial. She explains to the suspects that they each have two alternatives: they can confess to the crime, or not confess. If neither confesses, then she will charge them on some trivial offense, and their punishment will be relatively minor— one year in jail. If one of them confesses, then he will get off scot-free, while the other will serve ten years in prison. If both confess, then they will both be prosecuted, but the District Attorney will recommend a less severe sentence— eight years each.

What should the suspects do?

Put yourself in the shoes of one of the suspects. No matter what your partner in crime does, you are better off by confessing. (If your partner is silent, you will get off completely, and if your partner confesses, you will receive a sentence of eight instead of ten years.) However, if both suspects act in their individual self-interest and follow this logic, then both will confess and both will receive eight year sentences. This is much worse for them than if they had both remained silent. Ironically, the strategy which would serve the best interests of each individual fails to serve the best interests of either when applied by both individuals.

Game Theory

Another well-known idea with a game-theoretic flavor is the following: two distrustful people wish to share a pie equally. One way of doing this is to have one person cut the pie and let the second choose the piece. That way the cutter will try as much as possible to cut the pie fairly. (How could we generalize this procedure if there were three people rather than two?)

Game theory is also used to analyze things that we usually think of as "real" games. For example, the popular game *Connect-Four*™ was completely solved in 1984 — there is a perfect strategy for the first player. In other words, if you are the first player and you know the strategy, you can always win, no matter how brilliant your opponent is.

One way of analyzing certain games is to look at all possible ways in which the game can be played out. For example, with Tic-Tac-Toe one would realize that there are essentially only three opening moves:

In terms of strategy, a move to the lower right corner is really just the same as a move to the upper left; just consider the entire board rotated 180° about the center square. You will notice that the same principle of symmetry was invoked in *Two Theorems about Magic Squares* (p. 35).

Each of these first moves can be followed by only so many second moves. All possibilities can be explored and the best possibilities can be found. By this sort of analysis one can confirm the well-known fact that if both players play reasonably well, a game of Tic-Tac-Toe will end in a draw. (A lesser-known fact: in some sense the best opening move is the corner. If the second player does not play in the center, then the first player can force a win.)

The best way to get a feel for game theory is to play around with some games and see what makes them tick. Sometimes games with very simple rules can have some subtle and beautiful mathematics lurking in the background. The most famous example of this is the game of Nim which we investigate in the next few pages.

Game Theory

The Game of Nim

Nim is perhaps the best-known game for which winning strategies have been completely analyzed. The rules are simple, but the winning strategy is far from evident. In fact, the strategy is simple, unexpected, and has connections to other parts of mathematics.

How to Play

•Nim is played by two players with several rows of toothpicks.
•Each player alternates removing some toothpicks from a single row.
•The person who removes the last toothpick is the winner.

For example, a common starting configuration consists of five rows, with 1, 2, 3, 4, and 5 toothpicks respectively as shown below.

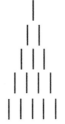

To whet your appetite
The winning strategy requires that the first player *must* remove exactly one toothpick from one of the odd rows, or the second player will be able to win.

If a game began with 3 rows as shown below, it might proceed as follows:

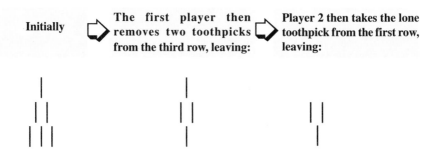

Initially ⟹ **The first player then removes two toothpicks from the third row, leaving:** ⟹ **Player 2 then takes the lone toothpick from the first row, leaving:**

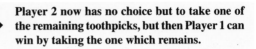

Player 1, sensing victory, takes a single toothpick from the middle row leaving:

Player 2 now has no choice but to take one of the remaining toothpicks, but then Player 1 can win by taking the one which remains.

$$|$$

$$|$$ $$|$$

*The winning strategy is explained in **How to Win at Nim** (p. 191). Before you read that section, you should find an opponent and play a few games of Nim, in order to get a feel for what makes a "good strategy", especially in the endgame when there aren't many toothpicks left.*

If you have enough self-control, you should try to figure out the winning strategy yourself, first in the small cases, and later in complete generality. This may mean pondering the game on and off for weeks or months. But the satisfaction in coming up with the winning strategy yourself is fantastic - it's no mean feat of ingenuity! If you're thinking of trying this route, sneak a peak at the Cryptic Nim Hint.

Cryptic Nim Hint

SBUIFS UIBO XSJUJOH EPXO UIF
TJAF PG UIF QJMFT JO CBTF 10,
XSJUF UIFN EPXO JO CBTF 2.

To decode the **Cryptic Nim Hint,** write the above statement replacing each letter with the letter which precedes it in the alphabet.

Game Theory

Game Theory and Politics: Arrow's Theorem

Philosophers and laypeople alike have debated the question, "What is the best form of government?" for millennia. In this day and age, our society has come to the consensus that democracy is in some sense the best. (Of course, this begs the question, "Whose version of democracy is the best?")

The quest for a "perfect" system of government can be viewed from a mathematical perspective. There are certain "irrationalities" in some of the candidates for the best system. For example, under the "majority rules" system, a decision between two choices is made simply by taking a vote. But something odd can happen when more than two choices are possible. For example, there might be four candidates for the presidency of a country. One of them leans (politically) to the right, and is popular with 28% of the voting public. Three of them lean to the left, and have similar platforms — they are each supported by 24% of the voters. Thus, a leader will be elected from the right, although the vast majority (72%) of the population would have preferred one from the left. Some people would just consider this "tough luck"; others would suggest that a slightly different system would be fairer.

Kenneth Arrow, an economist at Harvard University, addressed this issue in 1951. (He won a Nobel Prize for this and other work in 1972.) He started by demanding that a system satisfy the barest minimum standards of rationality.

We want some system to turn our individual desires into a collective choice. Imagine that we have a country (or state, or committee, or school) with three or more individuals, and we want to make a collective choice among several (more than two) options. Everyone writes down their preferences (an ordering of the options) on their ballot. These are then sent to a computer, and the computer produces a single result, listing the options in order of collective preference. The computer can use any crazy algorithm, subject only to the following minimal reasonable restrictions:

1. First of all, a non-restriction. We won't demand that all votes be counted equally. This allows the possibility of a system in which the value of a vote depends on the voter's age, education, profession, income, or eye color. For example, the computer may decide that college professors are out of touch with reality, and that their votes should be ignored.

Game Theory

2. If everyone prefers option X to option Y, then society as a whole will prefer option X to option Y. This is certainly a reasonable minimum condition!

3. (Independence of Irrelevant Alternatives) Society's decision between two choices should depend only on how individuals feel about those two choices. In terms of the 1992 U.S. Presidential election, this means that society's preference between Clinton and Bush should not be affected by whether Perot is on the ballot.

4. We'll assume that individuals aren't fools — they won't say that they prefer A to B, B to C, and C to A. So we will demand the same of society. Note that the "majority rules" system fails this condition. For example, consider a society with three people, with the following preferences:

Voter 1 prefers	A to B to C
Voter 2 prefers	B to C to A
Voter 3 prefers	C to A to B

Then if society operates by "majority rules", then society would prefer A to B, B to C, and C to A, each by a 2-1 vote!

The four conditions above seem like reasonable bare minimum conditions for an election. We have seen that "majority rules" fails this test, but one would think that surely some democratic system would pass. Not so, says Arrow:

Arrow's Remarkable Theorem

> The only system satisfying these minimal conditions
> is a dictatorship by one person.

This theorem has numerous disturbing implications, many of which should be immediately evident to the reader. One possible response is that many other factors need to be taken into account when judging a decision-making system, and although the varied democratic systems all fail one or another of the rationality conditions, they still work very well for other reasons. Readers might sympathize with Winston Churchill's well-known comment:

**Sir Winston Churchill
1874-1965**

Democracy is the worst system devised by the wit of man, except for all others.

Game Theory

Partial answers and discussion are given on page 72.

❶. There are certain paradoxes with a flavor similar to what we've seen here. Consider three (fair) dice, numbered unusually:

Die A: 1, 1, 5, 5, 5, 5
Die B: 3, 3, 3, 3, 3, 3
Die C: 2, 2, 2, 2, 6, 6

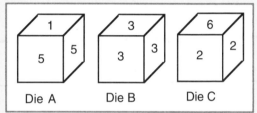

Two people play a game. They choose different dice. They then roll their respective dice, and the player with the higher number wins. If one player has A, and the other has B, it is fairly straightforward to see that A will win 2/3 of the time. B will always roll a 3, and A will beat that 2/3 of the time (by rolling a 5), and lose 1/3 of the time (by rolling a 1).

Who will most likely win in a competition between B and C? Who will most likely win in a competition between C and A? Why is that surprising?

If you are unsure how to proceed, take three normal dice, change their numbering, and experiment!

Game Theory

❷. In 1956, the U.S. House of Representatives voted on a bill calling for federal aid for school construction. An amendment was proposed stipulating that federal aid would be provided only to those states whose schools were integrated. The voters were loosely divided into three equal-sized interest groups: Republicans, northern Democrats, and southern Democrats. Of the three options available, the three groups had differing opinions. The Republicans favored no bill at all, but if one were to pass they preferred one restricting aid to integrated states. The southern Democrats, being from states with segregated schools, preferred the original bill first, no bill second, and the amended bill third. The northern Democrats preferred the amended bill, followed by the original bill and no bill. These preferences are summarized in a table.

Preference	Republicans	Southern Democrats	Northern Democrats
First	no bill	original bill	amended bill
Second	amended bill	no bill	original bill
Third	original bill	amended bill	no bill

Clearly, the original bill would have passed, as the Democrats together preferred the original bill to no bill at all. However, in keeping with House procedure, the first vote was on whether to accept the amendment; the Republicans and northern Democrats together ensured that the amendment passed, because they both preferred the amended bill to the original bill. The second vote was whether to approve the amended bill, or to have no bill at all. This time, the Republicans and the southern Democrats, both preferring to have no bill rather than the amended bill, conspired to ensure the bill's defeat. Paradoxically, the proposal of a popular amendment to a popular bill ensured the bill's eventual defeat! Analyze how this happened.

The author was first introduced to this topic by Paras Mehta of Oxford University. Arrow's Theorem is discussed further in Hoffman's *Archimedes' Revenge*. (See the *Annotated References* for bibliographic information.) The example of the federal bill for school construction is taken largely from that source.

Game Theory

Answers to Food for Thought (pp. 70-71)

1. Competition between A and B:

   B will always roll a 3. A will therefore win two-thirds of the time by rolling a 5 and will lose one-third of the time by rolling a 1. Thus A will usually beat B.

   Competition between B and C:

   By a similar argument to the one above, B will usually beat C.

   Competition between A and C:

   One method for determining who will win the competition between A and C is to list the possible outcomes in a table like the one below.

	A rolls a 1 (1/3 of the time)	A rolls a 5 (2/3 of the time)
C rolls a 2 (2/3 of the time)	A rolls a 1 and C rolls a 2 $\frac{2}{3} \times \frac{1}{3} = \frac{2}{9}$ of the time.	A rolls a 5 and C rolls a 2 $\frac{2}{3} \times \frac{2}{3} = \frac{4}{9}$ of the time.
C rolls a 6 (1/3 of the time)	A rolls a 1 and C rolls a 6 $\frac{1}{3} \times \frac{1}{3} = \frac{1}{9}$ of the time.	A rolls a 5 and C rolls a 6 $\frac{1}{3} \times \frac{2}{3} = \frac{2}{9}$ of the time.

From this table, it is clear that the probability that C will beat A is:

$$\frac{2}{9} + \frac{1}{9} + \frac{2}{9} = \frac{5}{9}$$

Incredible as it seems, we have just shown that A will usually beat B, B will usually beat C, and C will usually beat A!

2. This paradox is very similar to the paradox we encounter in the problem of the dice given above. It also relates to the failure of the "majority rules" system to satisfy condition 4 on page 69.

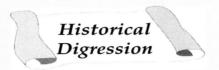

Historical Digression

Social Choice Theory

Social choice theory is a part of game theory that is studied by people in a number of different fields, including economics, political science, and philosopy. Its ties to political science relate to theories of voting, elections, and the state and its links to philosophy pertain to theories of equity, justice, and rational action. Although it is a young field, its roots lie as far back as the eighteenth century with Jeremy Bentham (philosophical foundations of utilitarianism and welfare analysis), John Charles de Borda, and the Marquis de Condorcet (theory of elections and committee decisions). Charles Lutwidge Dodgson, better known as Lewis Carroll, author of *Alice in Wonderland* and *Through the Looking Glass,* anticipated developments in social choice theory when, as an Oxford mathematics tutor, he made the first complete analysis of voting paradoxes.

THE BETTMANN ARCHIVE

**Lewis Carroll
1832-1898**

Kenneth Arrow established the current combinatorial framework of social choice theory. In *Archimedes' Revenge,* Paul Hoffman states:

> *The Nobel Prize-winning work of American economist Kenneth Arrow shows that achieving the ideals of a perfect democracy is a mathematical impossibility. Indeed, undesirable paradoxes can arise not only in voting but even before voting takes place, in deciding how many representatives are allocated to each district in a system of indirect representation, such as the House of Representatives* (p. 213). (See the *Annotated References* for bibliographic information.)

Are you a super sleuth? Try your hand at analyzing this game!

Game Theory

Elementary My Dear Watson!

S herlock Holmes, master of logical deduction, chases a hardened criminal through Europe. One player plays the dashing and daring Holmes, and the other player plays the evil, nefarious, dastardly, reprehensible, unscrupulous, notorious criminal— hereafter denoted by the acronym, ENDRUN.

The Game

This battle of wits can be played in the airports of Europe. A less expensive alternative is to use the map on the opposite page. Each of the dots represents a city where the chase can take place, and the players can indicate their position by pointing at a city with a pencil or a pen. The straight lines indicate air routes where the players may travel. Sherlock Holmes begins in Stockholm, where he begins his famous chase. The criminal, ENDRUN, starts off in Paris, where he has just stolen the Mona Lisa. Each player, in his or her turn, may jump from one city to another connected by an air link. Holmes, being true and noble, goes first. If Holmes lands in the same city as ENDRUN, he captures the thief after an exciting and excessively destructive car chase and wins the game. If Holmes does not catch ENDRUN in fifteen or fewer moves, ENDRUN escapes and wins the game.

The Problem

This game is fixed. If you pick the right side, and play perfectly, you can always win, no matter how well your opponent plays. The big question is: who has the winning strategy? (And what is it?)

A Hint

This is less a hint than plain common sense. Don't give up too soon! Try playing a few games with someone else, just to get a feel for how it works. Play both sides. Make a conjecture. Then try to figure out why it's right (or wrong!).

The answer is given on page 186.

Game Theory

Sherlock Holmes' Continental Chase

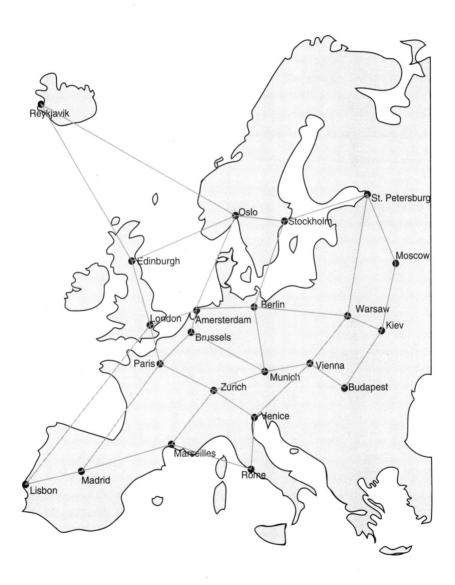

GEOMETRY

There is no royal road to Geometry.

— Menaechmus (to Alexander the Great)

The Mathematician Who Anticipated Calculus by Almost 2000 Years

Give me a place to stand on and I will move the earth.

Archimedes 287-212 B.C.

When historians of mathematics attempt to identify the greatest mathematician of all time, they inevitably produce a list of three giants: Archimedes, Newton, and Gauss. Although Archimedes lived in ancient Greece, his mathematical techniques were comparable in rigor to those of the greatest mathematicians of the 17th and 18th centuries. Among his many treatises was one titled *On Spirals* in which he presented twenty-eight propositions about spirals — including one which now bears his name.

His method of "exhaustion" anticipated the development of integral calculus by almost 2000 years! In his treatise *Quadrature of the Parabola*, he applied the method of exhaustion to derive the theorem presented on page 79, expressing the area of a segment of a parabola in terms of the area of a particular inscribed triangle. Archimedes was unable to find the area of a general segment of the other conics, the ellipse and the hyperbola, because these involve transcendental functions not known in his time.

Archimedes also made significant contributions to applied mathematics and engineering, including his famous law of hydrostatics. He developed a variety of practical devices such as the lever which he used to catapult huge quarter-ton rocks against an attacking Roman fleet during the seige of Syracuse in 212 B.C.

Of his countless contributions to many branches of mathematics, Archimedes was most proud of his discovery that the volume of a sphere is two-thirds the volume of the cylinder in which it is inscribed. (His method is described in *Archimedes Strikes Again* on page 82.) Archimedes requested that this discovery and the corresponding diagram be carved on his tomb. This request must have been granted, for we are told by the great Roman orator Cicero that he had restored Archimedes' tomb and the engraving when he was quaestor in Sicily.

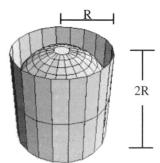

The volume of a sphere is two-thirds the volume of the right circular cylinder in which it is inscribed.

Geometry

Lengths, Areas, and Volumes
Arguments both True & False

Part 1: The Area of a Circle

The circumference C and area A of a circle with radius r can be calculated using these familiar formulas:
$$C = 2\pi r \quad \text{and} \quad A = \pi r^2$$

From these formulas, we can readily deduce the formula $A = \dfrac{rC}{2}$.

Here is another intuitive method to see why this is true. We divide the circle into n equal "pie slices" (no pun intended).

The full circle divided into n equal parts	One slice of the full circle is one of n equal parts

This entire argument can be better expressed in the language of limits. In fact, if you work out what is really going on, you'll rediscover the famous result:

$$\lim_{x \to \infty} x \sin\left(\frac{1}{x}\right) = 1 \text{ or } \lim_{x \to 0} \frac{\sin x}{x} = 1$$

This example is a great motivation for why we might expect the above limit to be true.

Each pie slice looks a lot like a triangle with base $\frac{C}{n}$ and height r (especially when n is very large). Then each triangle has area $\frac{rC}{2n}$. As there are n such pie slices, the total area of the circle is $n(\frac{rC}{2n}) = \frac{rC}{2}$.

Part 2: The Area of a Segment of a Parabola

Some readers may think that the foregoing argument is bogus. However, this is the method by which the ancient Greeks computed the area. Calculus of some sort is necessary to make this argument rigorous, but even without calculus the Greeks were able to prove some remarkable

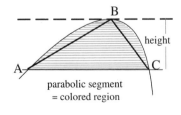

parabolic segment = colored region

results. For example, Archimedes wished to find the area of a *segment* of a parabola. (A parabola is a plane curve similar to the curve $y = ax^2$, for some real number a. A *segment* of a parabola is the closed region bounded by that parabola and any straight line.) Archimedes gave a rigorous proof that the area of the segment of the parabola is four thirds the area of a triangle having the same base and equal height.

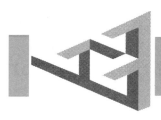

Geometry

Part 3: The Surface Area of a Sphere

You might still be feeling queasy about the lack of rigor in this style of argument —and rightly so. Here is a similar argument that gives a wrong answer. Can you find the flaw? (This is trickier than the usual "find-what's-wrong-with-this-proof" puzzles, as this argument doesn't claim to be rigorous in the first place.)

We divide the sphere of radius r into two hemispheres at its equator, and partition the equator into n segments, each of length $\frac{2\pi r}{n}$. We draw great circles through these partition points and the north and south poles of the sphere. Now imagine the sphere sliced along these great circles, peeled back and flattened out like a Mercator projection (for readers with a background in geography).

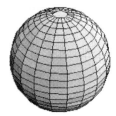 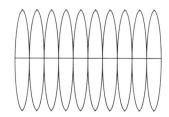

Each of these little triangles has base $\frac{2\pi r}{n}$ and height one-quarter the circumference of the sphere, $\frac{2\pi r}{4}$, and hence has area

$$\frac{1}{2}\left(\frac{2\pi r}{n}\right)\left(\frac{2\pi r}{4}\right) = \frac{\pi^2 r^2}{2n}$$

Since there are $2n$ triangles, the total area is $\pi^2 r^2$, which is the correct formula for the surface area of the sphere — NOT!!! Oh no! Some of you know that the surface area of a sphere of radius r is $4\pi r^2$, so we are off by a factor of $\frac{4}{\pi}$. What has gone wrong? How can the method that worked so well in Part 1 have failed us in Part 3? Philosophical pseudomathematics tells us that two wrongs do not make a right; but what do a right and a wrong make?

"Yes, yes, I *know* that, Sidney ... *everybody* knows *that* ... But look: Four wrongs *squared*, minus two wrongs to the fourth power, divided by this formula, do make a right."

Geometry

Part 4: The Volume of a Sphere

We can add to this confusing stew by working out the volume of the sphere. The reader should be able to show (by the same argument as in Part 1, this time by dissecting the sphere into n almost-pyramids with vertices at the center of the sphere) that if a sphere has surface area A and radius r, then the volume V of the sphere is given by $V = \frac{Ar}{3}$.

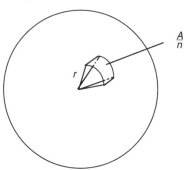

As the volume of the sphere is $V = \frac{4\pi r^3}{3}$ and the surface area is $A = 4\pi r^2$, the formula is correct. The method appears to be working again.

Food for Thought

❶. a) Define the "radius" r of a square to be the length of the altitude from the center of the square to one of its sides. Show that the formula from Part 1 ($A = rC$) still holds.
b) How should the "radius" of a regular n-gon be defined so the formula still holds?
c) What should the "radius" of a cube be for the formula in Part 4 to hold? How about a regular tetrahedron? Why are all of these answers similar?

❷. The "hypervolume" of a "four-dimensional sphere" of radius r is $H = \frac{\pi^2 r^4}{2}$. Can you use a method similar to that of Part 1 and Part 4 to find the "surface volume"?

Some of this material appeared in *Mathematical Mayhem* (Vol. 4, Issue 4, (March/April 1992)). Part of the erroneous argument is taken with permission from the *Mathematical Digest* (No. 86, (Jan. 1992)), 15. For more information on either of these publications, see the *Annotated References*.

Geometry

Archimedes Strikes Again

You've probably memorized the formula for the volume V of a sphere (in terms of the radius r: $V = \frac{4}{3}\pi r^3$), but you probably haven't seen why it is true. Here is an intuitive proof, originally by Archimedes.

First we recall one fact: the volume of a right circular cone with base of radius r and height

h is $V = \frac{1}{3}\pi r^2 h$.

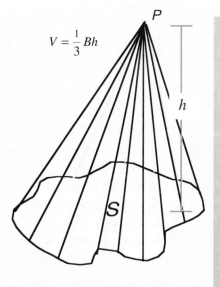

This is a special case of something more general. If you have a strange plane shape S with area B, and a point P at a height h above the plane, then the cone with base S and vertex P consists of everything that lies on a line segment with one endpoint in S and one endpoint at P. Then the volume of this cone is
$$V = \frac{1}{3}Bh .$$

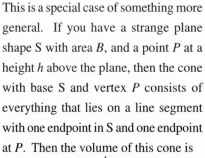

$$V = \frac{1}{3}Bh$$

This is a generalization of the fact that the area of a triangle with base b and height h is
$$A = \frac{1}{2}bh .$$

Can you propose the further generalization to four dimensions?

Archimedes Strikes Again (cont'd)

Archimedes computed the volume of a hemisphere follows. Place your hemisphere (of radius *r*) on a flat table A. Beside it, place a circular cone with base of radius *r*, and height *r*, balanced perfectly on its vertex.

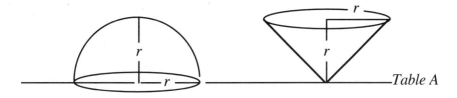

On another table (call it B), place a cylinder with base of radius *r*, and height *r*.

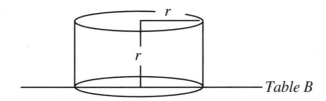

Now consider a cross-section over table A at height *h*.

As shown in the diagram, the cross-section of the sphere is a circle of radius $\sqrt{r^2-h^2}$.

The cross-section of the cone is a circle of radius *h*. (The proof uses the fact that the the shorter sides of a 90°-45°-45° triangle are equal.)

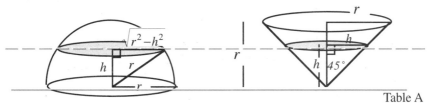

Table A

We are now in a position to execute Archimedes' beautiful heuristic proof.

83

Archimedes Strikes Again *(cont'd)*

Thus, the total cross-sectional area at height h above table A is

$$\pi\left(\sqrt{r^2-h^2}\right)^2 \quad + \quad \pi h^2 \quad = \quad \pi r^2$$
$$\qquad\uparrow \qquad\qquad\qquad \uparrow \qquad\qquad\qquad \uparrow$$
$$\text{hemisphere} \qquad \text{inverted cone} \qquad \text{total cross-section}$$

The cross-section of the cylinder is always a circle of radius r, so the total cross-sectional area above table B is also πr^2.

Here's the key step: for each height, the cross-sectional areas are the same, so the total volume must be the same. (You might want to ponder this for a while and convince yourself that it is reasonable.)

Thus: volume of hemisphere + volume of cone = volume of cylinder

so, volume of hemisphere + $\frac{1}{3}(\pi r^2)r$ = $(\pi r^2)r$

From this, we conclude that the volume of a hemisphere of radius r is $\frac{2}{3}\pi r^3$. Therefore the volume V of a sphere of radius r is given by:

$$V = \frac{4}{3}\pi r^3$$

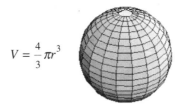

And the proof is complete! This clever argument surely deserves a shouted "Eureka", but I'm not sure if I'd run naked through the streets over it.

If you are comfortable with integral calculus, you will recognize how Archimedes' ideas anticipated its fundamental concepts. And if you haven't seen integral calculus before, keep this example in mind for when you do!

Geometry

The Falling Ladder Problem

❶ Late at night, a thief with a flashlight has climbed exactly halfway up a ladder in order to break into a house.

❷ Unfortunately (or fortunately, depending on your point of view), he hasn't secured the base of the ladder. (Don't try this at home!)

❸ The ladder slides out from under him, leaving him ignominiously on the ground.

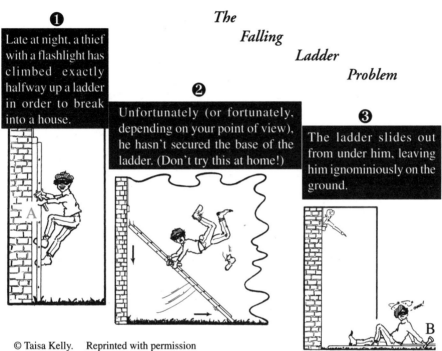

© Taisa Kelly.   Reprinted with permission

A police officer running to the scene observes the attempted burglary. Through the darkness, all she can see is the path of the flashlight. However, she is able to deduce from this the nature of the misadventure and later provide surprisingly accurate testimony at the burglar's trial.

What path did the flashlight take? Did it travel in a straight line between point A and point B? Before looking at the answer (in the *Solution to the Falling Ladder Problem,* p. 164), you may want to think about it, and come up with a guess of your own.

What would be the path of a point on the ladder but not at the half-way point?

Geometry

An Easy Proof that arctan 1/3 + arctan 1/2 = 45°

Here is a slick non-algebraic proof of the identity arctan 1/3 + arctan 1/2 = 45°.

Consider the triangle in the Cartesian plane with vertices at A (0,0), B(1,3), and C(2,1), and the point D(1,0).

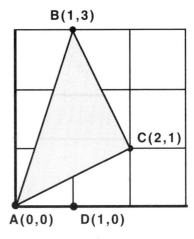

There are several ways of seeing that ABC is a 45°-45°-90° triangle. (For example,

$AC = BC = \dfrac{1}{\sqrt{2}} AB$.) Hence $\angle ABC = 45°$.

But $\angle ABC = \angle ABD + \angle DBC$.

Can you see why $\tan(\angle ABD) = 1/3$ and $\tan(\angle CBD) = 1/2$? If you can, then you've basically completed the proof!

The especially intrepid reader might want to try to find similar proofs of the following:

     a) arctan 1 + arctan 2 + arctan 3 = 180°.

     b) $\arctan \dfrac{1}{2} = \arctan \dfrac{1}{3} + \arctan \dfrac{1}{7}$

     c) More generally, if a and b are positive integers with $a > b$,

$$\arctan \dfrac{1}{a\text{-}b} = \arctan \dfrac{1}{a} + \arctan \dfrac{b}{a^2 - ab + 1}$$

*For another example of a proof by diagram, see **The Ailles Rectangle** on the opposite page.*[1]

---

[1]Many of these "proofs by diagram" have been collected in Nelsen's *Proofs Without Words*. See page 39 of Nelsen's book for a proof (due to Edward M. Harris) of the identity:
     arctan 1 + arctan 2 + arctan 3 = 180°.
(For bibliographic information, see the *Annotated References*.)

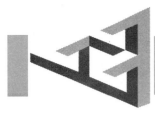
The Ailles Rectangle

Everyone who has learned trigonometry has memorized the sines, cosines, and tangents of "easy" angles such as 0°, 30°, 45°, 60°, and 90°. Doug Ailles, a high school teacher at Etobicoke Collegiate Institute in Etobicoke, Ontario, came up with an incredibly simple method of computing the trigonometric functions of 15° and 75°. In the long history of mathematical thought, it would not be surprising if someone had already invented a similar construction. But, in the absence of such information, I hereby christen it the *Ailles Rectangle*. Here's how it works.

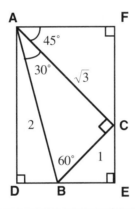

ABC is a 30°-60°-90° triangle with sides AB = 2, BC = 1, CA = $\sqrt{3}$. ABC is inscribed in rectangle ADEF such that ∠FAC = 45°.

Fill in the missing angles in the problem, and then (using the 45°-45°-90° triangles and the fact that AF = DE and AD = FE), find all missing lengths necessary to calculate the trigonometric functions of 15° and 75°.

The solution to this problem is given on the next page. Another way to compute these values is to substitute $x = 45°$ and $y = 30°$ into the sine, cosine, and tangent subtraction formulas:

$$\sin(x - y) = \sin x \cos y - \cos x \sin y$$
$$\cos(x - y) = \cos x \cos y + \sin x \sin y$$
$$\tan(x - y) = \frac{\tan x - \tan y}{1 + \tan x \tan y}$$

However, it is often more comforting to do things concretely and geometrically.

Geometry

Solution to the Ailles Rectangle Problem (p. 87)

Since the sum of the angle measures in a triangle is 180°, we can easily deduce that ACF and BCE are 45°-45°-90° triangles. Therefore AF = FC = $\frac{\sqrt{3}}{\sqrt{2}}$ and BE = CE = $\frac{1}{\sqrt{2}}$.

Using the above results with the equations AF = DE and AD = FE, we obtain the lengths shown in the figure below.

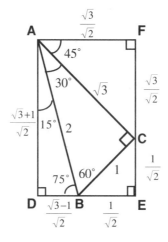

From triangle ABD in the figure, we can immediately write the following expressions for the trigonometric functions of 15° and 75°.

$$\sin 15° = \frac{\sqrt{3}-1}{2\sqrt{2}} \qquad\qquad \sin 75° = \frac{\sqrt{3}+1}{2\sqrt{2}}$$

$$\cos 15° = \frac{\sqrt{3}+1}{2\sqrt{2}} \qquad\qquad \cos 75° = \frac{\sqrt{3}-1}{2\sqrt{2}}$$

$$\tan 15° = 2 - \sqrt{3} \qquad\qquad \tan 75° = 2 + \sqrt{3}$$

Personal Profile

One of the Purest Forms of Mental Exercise

K a-Ping Yee was born in Toronto, Canada, but soon moved to Winnipeg, Manitoba, where he enrolled at St. John's-Ravenscourt, one of the top private schools in the country. Not long after arriving, Ping started to make a name for himself. Between grades four and six, he took the multiple-choice *Canadian National Math League* competition geared toward sixth graders; he achieved perfect scores in each of three consecutive years. John Barsby, a dedicated math teacher from the Upper School of SJR, noticed Ping and took him under his wing. He continued to work one-on-one with Ping throughout Ping's schooling.

Ka-Ping Yee (Canada)
Born February 26, 1976

In Grade 4, Ping also started doing Science Fair projects, later entering them in the Manitoba Schools' Science Symposium, the provincial science fair. In Grade 8, his project on chaos, fractals, and non-linear systems (an area of study just beginning to attract public attention) won him a place at the *Canada-Wide Science Fair*, as well as an appearance on the TV program *The Nature of Things*. The trip to the Canada-Wide Fair in St. John's, Newfoundland was his first solo trip away from home. He considers this experience a significant turning point in his life, opening his eyes to new people and new experiences.

In Grade 9, Ping again competed at the *Canada-Wide Science Fair*, this time with a project on evolving computer-simulated lifeforms. A prize there earned him a trip to the *London International Youth Science Fortnight* that summer. Once again he had a wonderful time, meeting people from over 50 other countries.

In Grade 10, besides winning another trip to the London Fortnight, he won the *Euclid* competition, a national competition for students two years his senior. Along with other successes, this achievement ensured him a place on the Canadian Team to the *International Mathematical Olympiad* (IMO) in Sigtuna, Sweden, where he earned an Honourable Mention.

89

The next year, in lieu of competing at the IMO, Ping decided to go to *Shad Valley*, an intense four-week program for gifted Canadian science students. In his final year of high school, he earned the highest score on both senior national mathematical competitions. These achievements won him a place at that year's IMO in Istanbul, Turkey, where he won a Gold Medal.

Meanwhile, Ping was also pursuing his interest in computers. In 1991 and 1992 he won the *International Computer Problem-Solving Contest* run by the University of Wisconsin, a contest in which competitors aimed to produce the highest calibre programs in the fastest possible time. After entering the University of Waterloo's competitive Computer Engineering program in the fall of 1993, he joined Waterloo's team to the *Association for Computing Machinery* programming contest. That team went on to win the international finals in Phoenix, Arizona, competing against undergraduate teams from across North America and elsewhere. In his first year at Waterloo, Ping also took part in the North American *Putnam Mathematical Competition* and received an Honourable Mention.

Ping has long had a fascination with the idea of applying pattern recognition to brainwaves, a process that with practice might enable us to communicate telepathically with computers and, to a certain extent, with other people. Nanotechnology is another fetish; he would love to explore the field of molecular manufacturing when he has the necessary background in chemistry and biology.

Ping's other current obsessions include Japanese animation (anime) and other things Japanese, graphic stories, devilsticks, and eighties synth-pop. He has also recently been attracted to a set of philosophies called "extropianism" that he discovered on the Internet. Seeking boundless expansion for the human race through technologies such as space travel, artificial intelligence, cryonics, and nanotechnology, extropianism questions traditional limits such as finite lifespans and rejects blind faith in favor of dynamic optimism. Ping finds that extropianism ties in well with his personal philosophy of getting involved with as many things and meeting as many people as possible.

Ping finds that his numerous successes all stem from his attitude toward life. His love of mathematics is one facet of an open attitude toward learning and experience: "Math is one of the purest forms of mental exercise — yet both the results and the methods of mathematics have applications everywhere. It is a great way to learn to solve problems, since you get volumes of essential knowledge at the same time. Math is a boundless universe that requires no more than an eager mind to explore, where you never have to worry about explosive decompression."

COMBINATORICS REVISITED

Strange as it may sound, the power of mathematics rests on its evasion of all unnecessary thought and on its wonderful saving of mental operations.

— Ernst Mach

Combinatorics Revisited

Parity Problems: The World Series

Everyone knows about odd and even numbers; there doesn't seem to be anything of great interest going on there. But in fact, parity (which means oddness or evenness), or more generally, symmetry, can be a powerful tool in solving problems. You may have noticed, for instance, that symmetry was the key principle in *A Mathematical Card Trick* (p. 44) and *The Water and Wine Puzzle* (p. 52) which appeared in the earlier chapter *Combinatorics*.

What follows in the next few pages is a selection of problems involving parity or symmetry in some way. I have started with a sample parity problem whose solution may help you with the remaining problems in this section Remember to play around with these problems for a while before you flip to the answers given later in the chapter!

Sample Parity Problem

Two teams, the Toronto Blue Jays and the Los Angeles Dodgers, are competing in the World Series. The two teams play games until one of them has won four games. (The series must end by the seventh game.) Each team (in this hypothetical universe) has a 50% chance of winning each game.

a)   What are the odds that the Toronto Blue Jays will win the World Series? (This is pretty easy!)

b)   Is it more likely that the World Series will end in six games or seven?

These don't look like your normal high school math problems! But they are in fact serious problems in combinatorics.

a) Your instincts should tell you immediately that there is a 50% chance of Toronto winning the series. After all, there is no difference between the teams. (Notice that our assumptions allow us to ignore all sorts of factors such as the quality of the teams, home field advantage, etc.) More explicitly, there is a symmetry between the two teams. If Toronto has probability p of winning the Series, then the Dodgers must win with probability p as well. As one of these two events must occur, each probability must be 50%. If you get jittery when you see a mathematical argument without equations, we can express this by saying that $2p = 1$, so $p = 0.5$.

Combinatorics Revisited

b) There are several ways of solving this problem. We will start with the more obvious one, and end with the more elegant. (The intrepid reader may want to try his or her hand at solving this problem before reading on.)

The "Brute Force" Solution

(If you find this method confusing, go on to the elegant solution.) We can compute the probabilities explicitly and compare them. To track the possible outcomes for the first three games, we can construct a tree diagram as follows:

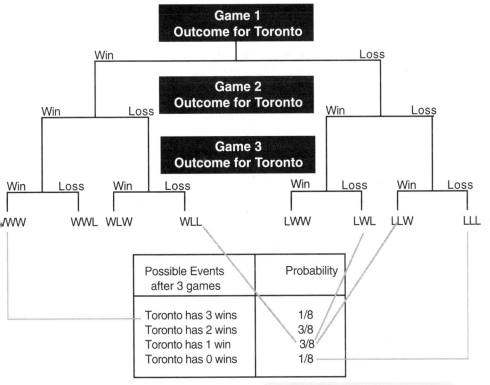

Tree Diagram of Possible Outcomes in the First 3 Games

Possible Events after 3 games	Probability
Toronto has 3 wins	1/8
Toronto has 2 wins	3/8
Toronto has 1 win	3/8
Toronto has 0 wins	1/8

Note: In the tree diagram above, two wins followed by a loss would be denoted WWL.

Notice that the numerators give the third row of Pascal's triangle, i.e. 1 3 3 1. Do you see why this is true?

93

Combinatorics Revisited

In each game the probability of winning is 1/2 for each team, so the probability of any particular outcome such as WWL is 1/8. We can extend the tree diagram above to compute the following probabilities for the possible situations after four or more games.

After four games

Possible Outcomes	Corresponding Events	Probability
WWWW	4 wins for Toronto	1/16
WWWL WWLW WLWW LWWW	3 wins for Toronto	4/16
WWLL WLWL WLLW LWWL LWLW LLWW	2 wins for Toronto	6/16
LLLW LLWL LWLL WLLL	1 win for Toronto	4/16
LLLL	0 wins for Toronto	1/16

The series terminates when either team wins 4 games, i.e. WWWW or LLLL. Hence the probabilities for the possible outcomes after the fifth game are:

After five games

Event	Probability
Series already finished after 4 games	1/8
4 wins for Toronto	4/32
3 wins for Toronto	10/32
2 wins for Toronto	10/32
1 win for Toronto	4/32

Series ends after 5 games.

Combinatorics Revisited

(We observe that the Pascal's triangle pattern observed after 3 games has mysteriously gone astray. What happened?)

Proceeding as above, we obtain the following probabilities for the possible events associated with the six and seventh games.

After six games

Event	Probability	
Series finished after exactly 4 games	1/8	
Series finished after exactly 5 games	1/4	
4 wins for Toronto	10/64	Series ends
3 wins for Toronto	20/64	after 6 games.
2 wins for Toronto	10/64	

After seven games

Event	Probability
Series finished after exactly 4 games	1/8
Series finished after exactly 5 games	1/4
Series finished after exactly 6 games	5/16
Series finished after exactly 7 games	5/16

We see that there is a 5/16 chance of the game finishing in six games, and a 5/16 chance of the game finishing in seven, so the probabilities are exactly the same. What a coincidence! (Or is it?)

The Elegant Solution

Notice that if, after five games, one team has won 3 games while the other has won 2, then the series must go to 6 or 7 games. Conversely, if the series goes to 6 or 7, then after 5 games one team must be leading 3-2. (If the score were 4 to 1, then one team would have already won; 5-0 is impossible, as the Series ends as soon as one team has won 4.) If the team that has won 3 of 5 wins Game 6, then the series ends at 6 games. There is a 50% chance of this. But if the other team wins Game 6, then the series is tied 3-3, and must go to the seventh game. There is a 50% chance of this too. (We don't even care who wins Game 7!)

You can see that the second solution has an edge over the first, although the "brute force" solution still works.

95

Combinatorics Revisited

More Parity Problems

The next few problems can be solved by using either "brute force" arguments or clever, more elegant techniques. As usual, the elegant solution is shorter and less painful to execute, although more difficult to conceive.

1. Imagine that the rules of the World Series have been changed, and that the Series continues until a team has won 11 games (so that it must end by the 21st game). Toronto and Los Angeles are in it again, and once again they are perfectly matched. Is it more likely that the series will end in 20 games or 21 games?

2. Let n denote a positive odd number. n pennies are flipped simultaneously, and the number of heads and tails turned up are counted. What is the probability that there are more heads than tails?

3. Let n denote a positive integer. You and I play a silly game. I flip n pennies, and you flip $n+1$ quarters simultaneously. You win if you flip more heads than I do; otherwise, I win. (If we flip the same number of heads, I win. In this way I hope to, at least partially, counteract the advantage you have of flipping an extra coin.) What are the odds of your winning?

The following problem is an entirely different kettle of fish:

4. Let n denote a positive odd integer. Imagine a table of numbers, with n rows and n columns. All of the entries are integers between 1 and n, and each integer between 1 and n appears exactly once in each row and column. Furthermore, the table is *symmetric about its main diagonal*; this means that the number appearing in the ith row and j^{th} column is the same as the number appearing in the jth row and i^{th} column. (See the diagram for an example.) Prove that each integer between 1 and n appears exactly once in the main diagonal. (The main diagonal consists of those numbers whose row number is the same as their column number.)

Sample Table for $n = 5$

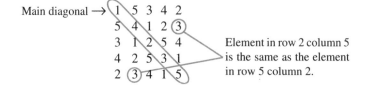

Combinatorics Revisited

More Parity Problems (cont'd)

5.   We play yet another arbitrary game.   You flip 30 pennies.   If the number of heads turned up is divisible by 3, you lose.   Otherwise, I win.   What are your odds of winning?

The *Solutions to the Parity Problems* appear on page 103. But don't read them until you have tried your hand at the problems! The solutions will be worth the wait. (If you're feeling pretty cocky and confident with parity, flip back to *Elementary My Dear Watson!* on page 74.)

Combinatorics Revisited

The Triangles of Pascal, Chu Shih-Chieh, and Sierpinski

Pascal's Triangle is a table of integers that arises in many different places in mathematics. Each row (from the second on) is obtained from the previous one by a simple rule. In each position, put the sum of the two numbers above it. For example, in the diagram, 10 is the sum of 4 and 6. The entries in row n are denoted

$$\binom{n}{0} \quad \binom{n}{1} \quad \binom{n}{2} \quad \cdots \quad \binom{n}{n}$$

so that, for example, the third element in row 4 is $\binom{4}{2} = 6$. $\binom{n}{k}$ is read as "n choose k".

We can calculate $\binom{n}{k}$ using the formula, $\binom{n}{k} = \dfrac{n!}{k!\,(n-k)!}$ where $n!$ denotes the product of the integers from 1 to n and $0! = 1$.

				1					row 0
			1		1				row 1
		1		2		1			row 2
	1		3		3		1		row 3
1		4		6		4		1	row 4

```
           1                       row 0
         1   1                     row 1
       1   2   1                   row 2
     1   3   3   1                 row 3
   1   4   6   4   1               row 4
 1   5  10  10   5   1             row 5
1  6  15  20  15   6   1           row 6
1 7  21  35  35  21  7  1          row 7
1 8 28  56  70  56  28  8  1       row 8
```

Although this triangle bears the name of Blaise Pascal (1623-1662), who came up with the idea when he was 13, it was known long before he was even born. The diagram on the next page, titled "The Old Method Chart of the Seven Multiplying Squares", appeared at the front of Chu Shih-Chieh's *Ssu Yuan Yii Chien* in 1303.[1]

---

[1] *A History of Mathematics*, by Boyer and Merzbach, (p. 206).

By looking at this table, and comparing it with Pascal's Triangle on the previous page, you should be able to work out the numbering system used in Chu Shih-Chieh's table.

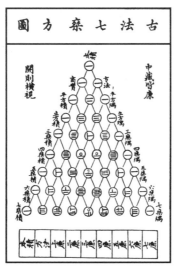

Some of the properties of Pascal's Triangle can be seen through another triangle consisting of A's and B's that is formed according to similar rules. Instead of 1's, write A's along the arms. Then fill in the AB-Triangle as follows. Each A has two different letters above it while the letters above each B are the same.

The **AB**-*Triangle*

		Number of A's per Row
row 0	A	1
row 1	A A	2
row 2	A B A	2
row 3	A A A A	4
row 4	A B B B A	2
row 5	A A B B A A	4
row 6	A B A B A B A	4

Do you see what the A's and B's correspond to in the original Pascal's triangle?

Combinatorics Revisited

Several things are immediately obvious about the AB-Triangle. First of all, it is symmetric. Second, there are B's along the center column, except for the top A. But there is a lot more going on too. Count the A's in each row. The first thing you'll notice is that this number is always a power of two. Why should this be true? Although the actual power of two seems to be quite random, there is a pattern. Write the integer n in base 2, and count the number of ones that appear. If k ones appear, then there will be 2^k A's in the nth row. For example, 5 is written as 101 in base 2, and there are 2^2 A's in row 5. (Using this, we can make statements such as "there are $2^5 = 32$ A's in the 361st row," because 361 written in base 2 is 101101001, which has five 1's.)

Now write down the AB-triangle, but leave out the B's. We first write down row 0, then the rows 0 and 1, then rows 0 to 3, and then rows 0 to 7.

Row 0

Rows 0 and 1

Rows 0 to 3

Rows 0 to 7

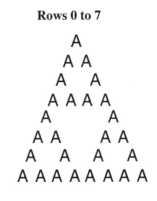

These figures are very similar to those that appear in the construction of a fractal called *Sierpinski's triangle*. (You may want to ponder why these two constructions generate similar diagrams.) Sierpinski's triangle is constructed as follows: Start with a solid equilateral triangle, divide it into four equal pieces as below, and remove the middle piece. Then divide each of the three remaining triangles into four equal pieces, and remove the middle piece of each one.

Step 1 **Step 2**

Step 3 **Step 4**

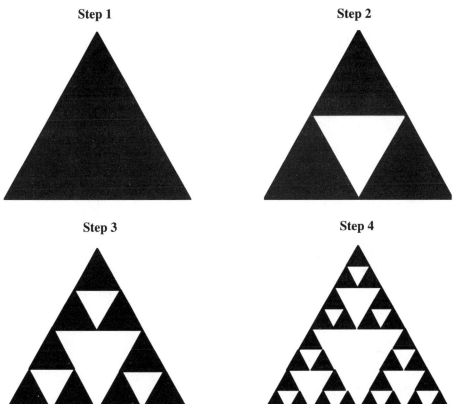

If this pattern were continued forever, then the result would be Sierpinski's triangle. If you made three copies of the triangle, and shrunk them down by a factor of one-half, then you could cover the original triangle. Because the figure has this *self-similarity* property, it is a fractal. (Because of its resemblance to Sierpinski's triangle, Pascal's triangle exhibits what can be loosely described as fractal-like behavior as well.)

Food for Thought

Many other surprising patterns turn up in Pascal's Triangle. You can discover some for yourself by playing with some of these ideas.

❶. Add up the elements in each row. What pattern do you notice? Now alternately add and subtract elements in each row. (For example, for the fourth row you would get 1 - 4 + 6 - 4 - 1 = 0.) What do you notice? (You might not find this surprising if you've seen some of the uses of Pascal's Triangle.) Can you prove any of your conjectures?

❷. What do you notice about the *parity* of $\binom{2n}{n}$? For which n is it odd, and for which is it even? Can you prove your conjecture?

❸. How does the parity of $\binom{2a}{2b}$ compare with the parity of $\binom{a}{b}$?

❹. Can you notice any other patterns in the AB-triangle? Make some conjectures, test them, and try to prove them.

❺. We've seen interesting things happening in mod 2. What about modulo other numbers? Mod 3 is one place to start. Let $a(n)$ be the number of elements in the nth row that are congruent to 1 mod 3, and let $b(n)$ be the number congruent to 2. What do you notice about $a(n)-b(n)$? Can you figure out the pattern? (Hint: look in base 3. It isn't easy!)

❻. What is the area of Sierpinski's triangle as a fraction of the area of the original triangle? If the perimeter of the original triangle is one unit, what is the length of the boundary of the Sierpinski triangle?

For more on modular arithmetic and different bases, flip back to **Divisibility Rules** *(p. 22). For a surprising connection between the Fibonacci numbers and Pascal's triangle, turn to* **Funny Fibonacci Facts** *(p. 146).*

Combinatorics Revisited

Solutions to More Parity Problems (p. 96)

1. Imagine that the rules of the World Series have been changed, and that the Series continues until a team has won 11 games (so that it must end by the 21st game). Toronto and Los Angeles are in it again, and once again they are perfectly matched. Is it more likely that the series will end in 20 games or 21 games?

> This problem is exactly the same as part b) of the Sample Parity Problem! Can you see why? (Of course, the numbers have been changed to protect the innocent.)

2. Let n denote a positive odd number. n pennies are flipped simultaneously, and the number of heads and tails turned up are counted. What are the odds that there are more heads than tails?[1]

> If you don't know where to start, borrow n cents from a friend and experiment for awhile. Start with $n = 1$, and then try $n = 3$. Try to guess what the answer should be based on your experimentation. If you're still stuck (which is likely), read the solution to the next problem and then come back. (Added bonus: When you've found a solution, ask yourself why you needed an odd number of pennies for it to work.)

3. Let n denote a positive integer. You and I play an arbitrary game. I flip n pennies, and you flip $n+1$ quarters simultaneously. You win if you flip more heads than I do; otherwise, I win. (If we flip the same number of heads, I win. In this way I hope to, at least partially, counteract the advantage you have of flipping an extra coin.) What are the odds of your winning?

> The solution to this problem is sneaky. The key observation is that, no matter what happens, either you flip more heads than I, or you flip more tails. Both can't happen, as you flip only one more coin than I. And neither can't happen either. (Think about it.) The odds of your flipping more heads is (by symmetry) just the same as the odds of you flipping more tails, so both probabilities are 50%. Thus, you will tend to win half the time. (You might want to experiment with this game.)

---

[1]Problem 2 has appeared before in many places, including the 1982 *Australian Mathematical Olympiad*.

Combinatorics Revisited

Solutions to More Parity Problems (cont'd)

Another elegant solution is to recognize that it doesn't matter whether you toss your $n + 1$ quarters simultaneously or whether you toss n of them and then toss the extra one. Suppose we each toss n coins simultaneously. If one of us tosses more heads, that person will win; the toss of the extra coin will not make any difference. Furthermore, since we each toss the same number of coins, we each have an equal chance of winning on this first toss. The toss of the extra coin will be needed (as a tiebreaker) only if we achieve the same number of heads on the toss of the n coins. If the extra coin comes up heads, then you win; otherwise we end up tied and I win. In either case, our chances of winning are equal.

4. Let n denote a positive odd integer. Imagine a table of numbers, with n rows and n columns. All of the entries are integers between 1 and n, and each integer between 1 and n appears exactly once in each row and column. Furthermore, the table is *symmetric about its main diagonal*; this means that the number appearing in the ith row and jth column is the same as the number appearing in the jth row and i^{th} column. Prove that each integer between 1 and n appears exactly once in the main diagonal. (The main diagonal consists of those numbers whose row number is the same as their column number.)[1]

Each number appears once in each row, and as there are n rows and n is odd, each number appears in the table an odd number of times. We can group those numbers not in the main diagonal into pairs using the symmetry property — pair the number in the ith row and j^{th} column with the (same) number in the jth row and i^{th} column. As each number appears an odd number of times, it must appear at least once in the diagonal. (Otherwise, it would appear in pairs, and thus would appear an even number of times.) Thus, we've just shown that each of n numbers appears at least once in the main diagonal, in which n numbers appear. Therefore, each of the n numbers must appear exactly once.

5. We play yet another arbitrary game. You flip 30 pennies. If the number of heads turned up is divisible by 3, you lose. Otherwise, I win. What are your odds of winning?

This is a trick question. If you read the question carefully, you'll see that you can never win!

---

[1]Problem 4 appeared in the 1954 *William Lowell Putnam Mathematical Competition.*

CHESSBOARD COLORING

The chessboard is the world, the pieces are the phenomena of the universe, the rules of the game are what we call the laws of Nature. The player on the other side is hidden from us. We know that his play is always fair, just, and patient. But also we know, to our cost, that he never overlooks a mistake, or makes the smallest allowance for ignorance.

Thomas Huxley

Chessboard Coloring

The Invention of Chess

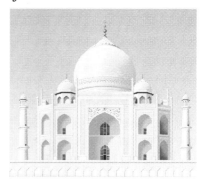

An old legend explains that chess was invented more than a thousand years ago for the Hindu King, Shirham, by his Grand Vizier, Sissa Ben Dahir. When offered a reward by the king, Sissa asked that he receive enough wheat to cover the squares of the chessboard in the following way. He wished to place a single grain on the first square of the board, two grains on the second square, four grains on the third square, and so on, placing twice as many grains on each succeeding square until all 64 squares were covered.

The king was astonished that Sissa would make such a modest request until he discovered that this would require more grain than there was in the entire world. Some accounts of this legend describe how the king was angered when he realized the enormity of the request and ordered that Sissa Ben Dahir be executed.

The total number of grains of wheat can be written as a geometric series with common ratio 2:

$$1 + 2 + 2^2 + 2^3 + 2^4 + \dots + 2^{63}$$

This sum can be found readily to be $2^{64} - 1$ by multiplying all terms by 2 and subtracting the original series from the resulting series. The sum $2^{64} - 1$ can be evaluated using a calculator:

$$2^{64} - 1 = 18\ 446\ 744\ 073\ 709\ 551\ 615$$

This example shows a geometric series with some particularly interesting properties. Observe, for example, that the sum of the first n terms of the series is one less than the $n + 1^{st}$ term.

The legend told above involved covering a chessboard with grains of wheat. In the remaining examples of this section, we will study the properties of various tilings of the chessboard. That is, we will look at coverings of all the squares of a chessboard with objects like dominoes such that every square is covered and none of the covering objects overlap.

106

Chessboard Coloring

A Chessboard Tiling Problem

Coloring doesn't refer to the use of crayons in mathematics, although coloring arguments are among the prettiest in elementary mathematics. Elegant and surprising, they are in retrospect easily believable, like a good whodunit. Coloring falls loosely into the category of combinatorics. Strictly speaking, coloring isn't a mathematical field of research, although it is a useful technique.

The best-known (and best-loved) introductory coloring problem is a chessboard tiling problem. You are given a chessboard. You are also given 32 dominoes, each consisting of two squares exactly the same size as the chessboard squares. (In case you are not a frequent chess or checkers player, the chessboard is an eight-by-eight board shown here.) The first problem is to *tile* the chessboard. This means that you want to cover the chessboard with dominoes, in such a way that each domino covers exactly two squares of the chessboard, and there is no overlap. This problem is ridiculously easy. The diagram shows two of the many solutions possible using 32 dominoes like this:

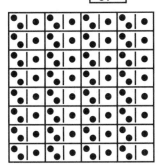

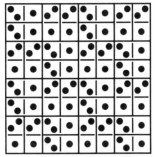

The next problem is more challenging. This time, cut out two opposite white corners of the chessboard. (Don't try this on the hand-carved chessboard from China!) Now, try to tile the new, mutilated board with 31 dominoes.

Can you tile this board with 31 dominoes?

When you are done (or have given up!), flip to A Solution to the Chessboard Tiling Problem *on the following page.*

Chessboard Coloring

A Solution to the Chessboard Tiling Problem

Before you read this section, you should first read

A Chessboard Tiling Problem

(p. 107).

This solution may seem like a sneaky trick (and in fact it is), but as you'll soon see, this trick can be used to solve a variety of seemingly impossible problems. The first step in solving the problem is to try to actually tile the chessboard with the dominoes. But after trying for several years without success (and losing many friends in the process), you will probably convince yourself that the tiling problem is impossible. Now comes the tricky part. It's easy to show that something can be tiled — all you have to do is exhibit a tiling. But how do you prove that something is untileable? You can't just try all possibilities — there are too many of them. So what can you do?

Well, the chessboard has some structure that you can use. Each square is black or white. If you think about it for a second, you'll see that no matter how you place a domino on the board so that it exactly covers two adjacent squares, it will cover one black square and one white square. Now, if you could tile the reduced chessboard with 31 dominoes, then exactly 31 white squares and 31 black squares must be covered. On the reduced chessboard there are 30 white squares and 32 black squares. (Remember that we removed two white squares from our chessboard!) Therefore, you couldn't tile the reduced chessboard with 31 dominoes.

This method of proof is called the *indirect method*. It's an excellent way of proving that something is impossible. You assume that it is in fact possible, and deduce a contradiction. Since the reasoning was flawless, the only possible explanation for this contradiction is that the original assumption was false. The best way to understand how this method works is to see it in action a few times. You've already seen it once here; it appears again in *Prime Numbers in Number Theory* (p. 124) and *The Harmonic Series* (p. 179).

If you're still leery of this argument, you could state it in another way. Each domino you place on the board must cover exactly one white square. As there is no overlap, each white square must be covered by at most one domino. So, as there are 30 white squares, you can't fit more than 30 dominoes on the reduced board, which isn't enough to cover the entire board.

Chessboard Coloring

If you're still not convinced, draw a little chessboard that is two squares by two squares, and another little chessboard that is four squares by four squares. Cut off two opposite white corners of each, and try to cover them with dominoes. You'll soon see intuitively that you have too few white squares to tile the boards.

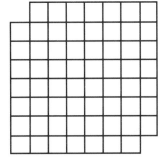

Next, let's solve a (very slightly) different problem. Take a new chessboard, and paint all of its black squares white. Now we have an 8 × 8 board where all of the squares are white. Cut out opposite corners. Can you tile this reduced board with 31 dominoes, if (as before) each domino must fully cover two squares of the reduced board?

"Of course not!" you bellow. "Painting the black squares white doesn't change the argument one bit!" Correct. We begin our argument with the statement, "Paint alternate squares black and white, as on a chessboard," and then continue as before.

That's what coloring problems are all about. You have to figure out a clever coloring of the object you're looking at, and make an argument from there. In *Another Chessboard Tiling Problem* (p. 111), you'll see some of the power of this technique.

Food for Thought

If you haven't had your fill of domino tiling, you can try the following problems:

❶. You're given a 100 × 100 board with two opposite corners removed. Can you tile the rest of the board with 4999 dominoes?

❷. (This is trickier.) I take a new chessboard, and using an Exacto knife, I remove two squares from the board, one of each colour. Can I tile the remaining 62 squares with 31 dominoes? Prove it.

Notice that you don't know which two squares I've removed, other than that one is black and one is white. Hint: the answer is always yes.

109

A Tetromino Tiling Problem

A tetromino consists of four squares, glued together along edges. There are essentially 5 tetrominoes:

although you can get more if you rotate them and flip them.

> The tetrominoes above have total area 20, so a natural question to ask is: Can you tile a 4 × 5 board using exactly one copy of each of these tetrominoes? (An easier question is: Can you tile a 2 × 10 board with them? And the easiest is: Can you tile a 1 × 20 board with them?)

Try to solve this problem yourself, by coloring the 4 × 5 board in the right way, and making the appropriate argument. Once you solve it (or once you give up), you can read *The Solution to the Tetromino Tiling Problem* (p. 113).

Another Chessboard Tiling Problem

Here is a harder puzzle than the previous tiling problems. Instead of dominoes or tetrominoes, you have 21 pieces that look like this:

and one piece that looks like this:

We'll refer to these pieces as 1 × 3's and 1 × 1's, for obvious reasons.

Now imagine a chessboard with squares just the size of the 1 × 1. You will notice that all of our pieces combined consist of 21 × 3 + 1 = 64 squares. This is precisely the number of squares on the chessboard. Is it possible to cover the chessboard with these 22 pieces? Once again, there can't be any overlap, and the squares on each piece must cover completely the squares on the chessboard.

To answer this question, we bring out some more paint and paint the squares white, green and black as shown in the diagram. You'll notice that there are 22 green squares, 21 white squares, and 21 black squares. You'll also notice that, no matter how you place a 1 × 3 tile on the board, it will cover exactly one green square, one black square, and one white square. When you place all 21 1 × 3 tiles on the board, 21 squares of each color will be covered. Thus the 1 × 1 tile must be placed on a green square.

We take out our paintbrushes again and paint the chessboard in a different way as shown in this diagram We can make the same argument, and prove that the 1 × 1 tile must, once again, be placed on a green square.

> Therefore if a tiling by the 22 tiles exists, then the 1 × 1 tile must be located on a green square in *both* diagrams.

The only locations which appear as green squares in both diagrams are shown in the diagram. Therefore, if a tiling does exist using the twenty-one 1×3 tiles and the 1×1 tile, the 1×1 tile must be covering one of the green squares in the diagram.

You might have observed that these locations are symmetric: a rotation of 90° about the center of the grid maps each of these four green squares onto another. This leads us to an even more elegant way to deduce this result.

If we have a tiling of the chessboard, then it remains a tiling if the chessboard is rotated through 90°, 180° or 270°. Therefore the 1×1 square must be a green square which maps onto another green square under these rotations. The only such green squares are those four in the diagram above.

So we're done? Not yet...we've only shown that *if* a tiling exists, then the 1×1 tile must occupy one of the four positions shown above. It remains to demonstrate such a tiling.

1 Verify that if the 1×1 is in one of these positions, you can actually tile the remainder of the chessboard with the twenty-one 1×3's.

So in this case, a tiling *is* possible, but the coloring was still useful. With what we learned, we could tile the board very quickly. Without the "tip" as to where to place the 1×1, we could have been at it for hours before coming up with the correct tiling.

2 Try to solve the same problem with a 4×4 board (and five 1×3's and a 1×1) and with a 5×5 board (and eight 1×3's and a 1×1).

Solution to the Tetromino Tiling Problem (p. 110)

Color the squares black and white, like a chessboard.

There are 10 white squares and 10 black squares. The first four tetrominoes shown below must each cover two squares of each color. The last tetromino must cover 3 squares of one color and one square of the other color.

The five tetrominoes

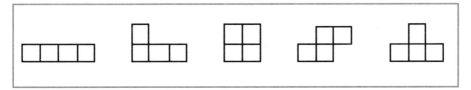

Assume that you can tile the 4 ×5 board with the five tetrominoes. Then the first four must cover 8 black squares and 8 white squares. This leaves 2 black squares and 2 white squares for the fifth one, which is impossible. Therefore, it is impossible to tile the 4 × 5 board as desired. (As a further challenge, you may want to replace the fifth tetromino with another one of the others and try to tile the 4 × 5 board.)

Personal Profile

A Sudden Flash of Insight

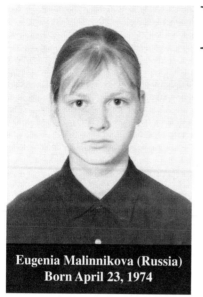

Eugenia Malinnikova (Russia)
Born April 23, 1974

Born and raised in Leningrad (now St.Petersburg), Eugenia Malinnikova has been interested in mathematics since early childhood. One formative mathematical memory dates from when she was twelve. Her father explained why the set of natural numbers and the set of real numbers between 0 and 1 do not have the same cardinality. (One version of this fact is discussed in *Two Different Infinities*, p. 229). She remembers being able to intellectually understand every word he said, but she "couldn't believe it or feel it."

One of the reasons for Soviet pre-eminence in mathematics was the presence of gifted mentors who guided young people into the mysteries of mathematics early in life. Serge Rukshin, a Leningrad mathematician with a long-standing interest in developing young talent, ran special "studying circles" for students from fifth to tenth grades (roughly ages 12 to 17). The students would learn math with Rukshin, but they also went to theaters, listened to classical music, and went for weekend walks outside of the city.

Eugenia was invited to join Rukshin's circle, along with a number of other young Leningrad students. Six members of the circle eventually went on to compete at the *International Mathematical Olympiad* (IMO), three for the USSR, two for the USA, and one for Israel. Today, all of them are continuing their studies in mathematics.

One of the first great insights Eugenia had at the studying circle concerned the following problem:

> Twenty-five jealous people live in the unit squares of a 5x5 grid. Each of them thinks that his neighbours in adjacent squares (horizontally and vertically) all live better than he does. Is it possible for all of them to move in such a way that everyone ends up in the square of one of his former neighbors?"

Eugenia tried to solve this problem at home for a few days, but to no avail. At the next lesson, the 5 × 5 square was drawn on the blackboard with squares painted in the pattern of a chessboard. A sudden flash of insight over came her and left her speechless.

What was Eugenia's insight?

Most of Rukshin's studying circle went on to Leningrad's School 239 in order to complete their final four years of high school, grades seven through ten. Bypassing the school's strict entrance exam, they took specialized courses in mathematics and physics. At the same time, Eugenia took courses in literature and found them to be among her favorite.

Starting in sixth grade, Eugenia took part in the *Leningrad Mathematical Olympiad*. Then, in March 1988, during seventh grade, she won first prize in the eighth grade competition, qualifying for the national olympiad. She won first prize there, the first step to making the Soviet Team to the *International Mathematical Olympiad*. The next year, she made the national team for the first of three times she would compete internationally. Although two years younger than the rest of the team, Eugenia went on to win a Gold Medal at the 1989 IMO in Braunschweig, Germany, losing only one mark on the entire competition. In her next two years, she won two more Gold Medals, achieving a perfect score each time. Such a performance is truly a rare occurrence at the Olympiad.

Since 1991, Eugenia has been studying mathematical analysis at St. Petersburg University. After her five-year undergraduate program, she will probably study for three more years to obtain a doctorate, although, as she acknowledges, "anything can happen". One fact is certain: she will always find mathematics "one of the most beautiful things in the world" .

NUMBER THEORY REVISITED

*A mathematician, like a poet or painter, is a maker
of patterns. If his patterns are more permanent than
theirs it is because they are made with ideas...A
mathematician...has no material to work with but
ideas, and so his patterns are likely to last longer.*

— G. H. Hardy

Numbers, Numbers, and More Numbers!

Jime out for some terminology

Throughout the next half of this book, you will be increasingly exposed to numbers that mathematicians commonly use. You will find integers, rational numbers, irrational numbers, complex numbers, algebraic numbers, transcendental numbers, and constructible numbers. The following pages will help you keep track of these concepts.

Real numbers are those numbers which can represent the length of a line segment.

Rational numbers are those numbers that can be expressed as a ratio of two integers (e.g. 42 or 2/3). *Irrational numbers* are those real numbers that aren't rational. In the next section, for instance, you will see that $\sqrt{2}$ is irrational, and in *Two Different Infinities* (p. 229) you will see that there are, in some sense, far more irrationals than rationals.

Complex numbers are extremely important in mathematics and physics. A complex number is any number of the form $a + bi$ where a and b are real numbers and i is an "imaginary" term satisfying the equation $i^2 = -1$. If you are unfamiliar with complex numbers, you should ask someone, such as a teacher, how they work; they are fascinating. Complex numbers will pop up at many points later on in this book; surprisingly, they often appear in geometric contexts.

Algebraic numbers are those complex numbers that are zeroes of polynomials with integer coefficients. For example, $\sqrt[3]{2}$ is algebraic, as $z = \sqrt[3]{2}$ is a solution of $z^3 - 2 = 0$.

Transcendental numbers are those complex numbers that are not algebraic. The mathematical constants π and e are transcendental, although the proofs of these facts require some advanced mathematical techniques. In the next section, you will see that $\sqrt{2}^{\sqrt{2}}$ is also transcendental. In the final section of the book, you will see that, in some sense, the vast majority of numbers are transcendental.

Constructible numbers are a special subset of algebraic numbers that are used to prove certain facts in geometry, including the fact that it is impossible to trisect an arbitrary angle using only a straightedge (unmarked ruler) and compasses. This will be further discussed in *Three Impossible Problems of Antiquity* (p. 215).

The following figure will help you keep track of the relationships among these concepts. For example, all rational numbers are algebraic, but not all algebraic numbers are real ($\sqrt[3]{2}\,i$ is an example).

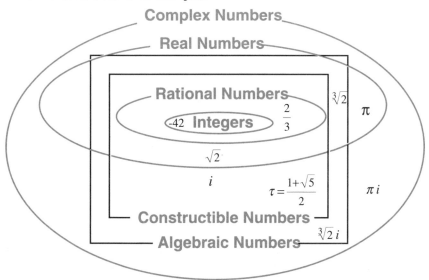

An Informal Glossary

•Integer	— any number in the set { ..., -3, -2, -1, 0, 1, 2, 3, ...} — an integer can be expressed as a sum or difference of two natural numbers
•Rational Number	— any number which can be expressed as a quotient of two integers
•Real Number	— any number which can represent the length of a line segment
•Complex Number	— any number of the form $a + bi$ where a and b are real numbers and $i^2 = -1$
•Constructible Number	— any complex number corresponding to a vector constructed in a finite number of steps from a unit vector by straightedge and compasses (see p. 215)
•Algebraic Number	— any complex number which is the root of some algebraic equation with integer coefficients

119

First Steps in Number Theory
Rational & Irrational Numbers

Number theory is the field of mathematics that deals with questions about the integers.[1] For example, Fermat's Last Theorem is a statement in number theory. (For more on this celebrated result, see the *Historical Digression* at the end of this section.) One aspect of the study of integers involves questions about rational and irrational numbers. Rational numbers are those real numbers that can be expressed as a ratio of two integers, such as 3/5 or 4. The rest are irrational numbers. Irrational numbers are harder to come by, although we'll see later (*Two Different Infinities*, p. 229) that in some sense there are far more of them. The mathematical constants π and e are irrational, although the proofs of these facts are quite difficult.

In the next few pages, we will demonstrate the power of the indirect method of proof in establishing two of the most beautiful theorems in mathematics. (You may already have encountered the indirect method in *A Solution to The Chessboard Tiling Problem*, p. 108.) In *A Mathematician's Apology*, the famous English mathematician G. H. Hardy (1877-1947) presents two theorems as examples of "first-rate" theorems in mathematics. In 1940, he wrote:

> "… I must produce examples of theorems which every mathematician will admit to be first-rate.…I can hardly do better than go back to the Greeks. I will state and prove two of the famous theorems of Greek mathematics.…Each is as fresh and significant as when it was discovered — two thousand years have not written a wrinkle on either of them (p. 91).
>
> …The proof is by *reductio ad absurdum* [the indirect method], and *reductio ad absurdum*, which Euclid loved so much, is one of a mathematician's finest weapons. It is a far finer gambit than any chess gambit: a chess player may offer the sacrifice of a pawn or even a piece, but a mathematician offers *the game* (p. 94)."

---

[1] Number theory actually goes much farther than this, but the integers are where number theory begins.

Number Theory Revisited

The following theorem is the first of the two referred to by Hardy. It dates back about 2500 years to the Pythagoreans, and is one of the simplest proofs that a number is irrational. It is also extremely beautiful — its elegance lies in its simplicity.

Theorem

$\sqrt{2}$ is irrational.

Proof.

This is an example of the *indirect method* of proof, in which we assume the opposite of the intended result. We then reach a contradiction. The only possible explanation for the contradiction is that our original assumption must have been flawed. Hence the intended result must be true. (If you find that confusing, just read on. It is really just common sense; the best way to understand the indirect method is to see it in action.)

What would happen if $\sqrt{2}$ were rational? Then we could write $\sqrt{2}$ as a fraction in lowest terms. Thus $\sqrt{2} = \frac{m}{n}$ where m and n are positive integers, and m and n have no common factors. In particular, m and n cannot *both* be divisible by 2.

$$\text{Then } \sqrt{2}\,n = m, \text{ so } 2n^2 = m^2.$$

The left side is even, so the right side must be too, so m is even. Therefore, $m = 2d$ for some integer d. Substituting this into our equation, we obtain:

$$2n^2 = (2d)^2 \Rightarrow n^2 = 2d^2$$

Now the right side is even, so the left side must be too, so n is even.

But we earlier said that m and n could not both be even! We have arrived at a contradiction, so there must have been a flaw in our original assumption that $\sqrt{2}$ is rational. Thus $\sqrt{2}$ must be irrational. Q. E. D.

(Can you prove that $\sqrt{3}$ is irrational by a similar method? $\sqrt{12}$? $\sqrt[3]{7}$?)

We now have an example that shows that a rational number taken to a rational power can be irrational. Can we take an irrational number to an irrational power and get a rational result? The next result answers the question in the affirmative, but doesn't actually give an example.

Theorem

> There exist two irrational numbers so that when the first is raised to the power of the second, the result is rational.

Proof.

We have already proven that $\sqrt{2}$ is irrational. If $\sqrt{2}^{\sqrt{2}}$ is rational, then we have found an example. Otherwise, $x = \sqrt{2}^{\sqrt{2}}$ is irrational, and

$$x^{\sqrt{2}} = \left(\sqrt{2}^{\sqrt{2}} \right)^{\sqrt{2}}$$

$$= \sqrt{2}^{2}$$

$$= 2$$

is rational, so we have found our example anyway.   Q. E. D.

(Can you find a similar proof using $\sqrt[3]{3}$?)

Oddly enough this proof doesn't actually resolve whether $\sqrt{2}^{\sqrt{2}}$ is rational! It turns out (although it is much, much more difficult to prove) that $\sqrt{2}^{\sqrt{2}}$ is irrational, and in fact transcendental.[1] This follows from the Gelfond-Schneider theorem:

The Gelfond-Schneider Theorem (1934)

> If a and b are algebraic numbers (i.e. the roots of polynomials with integer co-efficients), $a \neq 0$ or 1, and b is irrational, then a^b is transcendental.

In 1900, at the *International Congress of Mathematicians* in Paris, the famed German mathematician David Hilbert proposed 23 problems for the new century, including the *Riemann Hypothesis* (referred to in the profile of Noam Elkies, p. 241), the *Gelfond-Schneider Theorem*, and (in some sense) *Fermat's Last Theorem*. These problems inspired a tremendous amount of research, and most of them have now been solved. Perhaps another great mathematician will propose a new collection of problems at the 2000 *International Congress*, heralding a new century of progress.

---

[1]For a definition of transcendental, see the previous section, *Numbers, Numbers, and More Numbers!*

Historical Digression

The Most Famous Conjecture in Mathematics

Pierre de Fermat has been called the founder of modern Number Theory, although he is also known for his contributions to probability theory and analysis. While earning his living as a practicing lawyer in France, he pursued the study of mathematics as a hobby. *Fermat's Last Theorem* (stated below) was actually a conjecture scrawled in the margin of one of Fermat's books dating from about 1637. The challenge to prove or disprove Fermat's Last Theorem was enhanced by rewards offered for its solution. In 1815 and 1860, the French Academy of Sciences offered a gold medal and 300 francs for a proof, and in 1908, the German Academy of Sciences

Pierre de Fermat 1601-1665

offered a prize of 100 000 marks — a fortune in pre-inflationary Germany! In spite of these inducements, the conjecture defied proof until 1993.

Fermat's Last Theorem

If $n > 2$, then there is no solution to the equation $x^n + y^n = z^n$ with x, y, z positive integers.

On June 23, 1993, Andrew Wiles of Princeton University completed his final lecture at a Cambridge University conference with the statement, "I better stop there." The audience of some of the world's greatest number theorists was incredulous — paralysed for a few seconds. As the significance of the chalkboard equation registered, the silence gave way to a thunderous applause. It appeared that Dr. Wiles had proved Fermat's Last Theorem, the most famous conjecture in mathematics which had eluded the greatest minds for over 350 years!

Number Theory Revisited

Prime Numbers in Number Theory

As discussed in *First Steps in Number Theory* (p. 120), number theory is the field of mathematics that deals with questions about the integers. When one tries to prove results about integers, one naturally encounters the concept of a *prime number*.

We all know what a prime is: it is a positive integer whose only factors are 1 and itself. (In particular, 1 is not prime.) The first few primes are 2, 3, 5, 7, and 11. Early mathematicians asked themselves the reasonable question, "Is the list of primes finite?" In other words, "Is there a largest prime?" The mathematician Euclid of Alexandria (in present-day Egypt) had a very simple proof (c. 300 B.C.) that there are an infinite number of primes. It is a testament to his brilliance that he was able to come up with such a beautiful argument that it remains the best proof known to this day. This is the second theorem to which Hardy referred; like the first, this theorem uses the *indirect method of proof* discussed in the previous section.

Theorem

> There are an infinite number of primes.

Proof.

What would happen if there were only a finite number of primes? Well, in that case, we could call them $p_1, p_2, ..., p_n$ where all primes would be included in the list. Now consider the number

$$q = p_1 \, p_2 \cdots p_n + 1$$

Since q leaves a remainder of 1 when divided by any prime p_i, q is not divisible by any of the primes $p_1, p_2, ..., p_n$. In other words, it is not divisible by any prime. But this contradicts the fact that q *must* be divisible by *some* prime. We can't have contradictions in mathematics (which after all is a very serious subject), so there must have been a flaw in our reasoning somewhere. The only possible flaw is our assumption that there are a finite number of primes. So the number of primes must be infinite. Q. E. D.

The proof above gives a method of constructing primes. Given a set of primes, you can construct another one — take the product of the primes in the set, add 1, and take the smallest prime factor of the result. Starting with the set $p_1 = 2$, we get $p_2 = 3$, $p_3 = 7$, $p_4 = 43$.

Unfortunately, it is not a very efficient method of generating primes. It would be convenient if there were some polynomial that generated all of the primes in order. That's asking a little much, but it would still be convenient if there were some polynomial that took on only prime values. (Polynomials such as the constant 17 don't count!) The polynomial, $P(n) = n^2 - 81n + 1681$ takes on prime values when $n = 1, 2, 3, ..., 80$, but unfortunately $P(81) = 1681 = 41^2$ (and $P(0) = 1681$ too). Should we keep looking for such a polynomial? Probably not, in light of the following theorem:

Theorem

> There is no non-constant polynomial with integer coefficients that takes on only prime values.

Proof.

We will use the indirect method again. Assume such a polynomial, $P(n)$, exists. Then $P(0)$ is some prime p. This means that the constant term of the polynomial is p, so $P(n)$ is of the form $nQ(n) + p$, for some polynomial $Q(n)$ which also has integer coefficients. Then,

$$P(kp) = kp \, Q(kp) + p \text{ is divisible by } p, \text{ for any integer } k.$$

But by our hypothesis, $P(kp)$ is a prime. But the only prime divisible by p is p itself, so $P(kp) = p$ for all integers k. From here, you can finish off the proof in two ways, depending on your personal style.

Method I	*Method II*
Any non-constant polynomial $P(n)$ must grow indefinitely large as n gets large. This contradicts the fact that $P(kp) = p$ for all integers k.	$p = P(kp) = kp \, Q(kp) + p$, so when k is a non-zero integer, $Q(kp) = 0$. But $Q(n)$ is a non-zero polynomial, and as such can only have a finite number of zeroes, contradicting the fact that $Q(n) = 0$ for $n = ..., -3p, -2p, -p, p, 2p, 3p, ...$

Using either method, we have arrived at a contradiction, so our original assumption must have been incorrect. Thus no such polynomial exists.

We'll end this section with a simple test for primality. Suppose we have a number n, such as 101, and we want to check if it is prime. One way of doing this is to see if it has any divisors greater than 1 and less than n. Thanks to the next theorem, we can make our check much faster.

Theorem

> If n is composite (i.e. not prime), then n is divisible by a prime no greater than $\sqrt{n}$.

Proof.

If n is composite, then it can be expressed as $n = pq$, where p and q are greater than 1. Both p and q can't be bigger than $\sqrt{n}$ (or else $n = pq > \sqrt{n} \times \sqrt{n} = n$). Suppose $p \leq \sqrt{n}$. Then if p isn't prime, take any prime factor; this also will be no greater than $\sqrt{n}$.

Using this theorem, we can quickly check if 101 is prime. The prime numbers less than $\sqrt{101}$ are 2, 3, 5, and 7, and 101 is divisible by none of these, so it must be prime. (This theorem, along with the tricks mentioned in *Divisibility Rules* on page 22, allow one to check primality of reasonably large numbers in one's head.)

This method is quite crude, and extremely slow for checking the primality of integers that are many digits long. Much faster methods have been developed using sophisticated tools of number theory. Far from being arbitrary mathematical curiosities, these ideas are essential in cryptography and the creation of "unbreakable" codes.

The Painted Lockers

The lockers in Cretin High School are numbered 1 through 1000. Cretin's school colors are a fashionable mauve and vermilion. All of the lockers are initially painted mauve. After an excessively enthusiastic pep rally, Cretin's 1000 students go on a painting rampage. The first student runs down the halls and paints all of the lockers vermilion. The second student runs down the halls and repaints all of the even-numbered lockers mauve. The third student runs down the halls and repaints lockers 3, 6, 9, ..., 999 in the opposite color that he finds them. The fourth student runs down the halls and does the same with those lockers whose numbers are multiples of 4. This goes on and on and on, until finally (late that night), the thousandth student repaints locker 1000. After that peppy day, which lockers are painted mauve, and which lockers are painted vermilion? (Hint: imagine that the school only has 10 lockers, and work out what happens in that case. Then try it with 20 lockers. Then make a conjecture for the general case. Finally, try to prove it.)

© Taisa Kelly   Reprinted with permission

The answer is in *Solution to the Painted Lockers Problem* (p. 134).

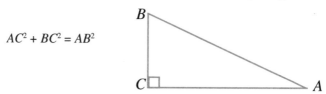

Number Theory Revisited

Geometry Meets Number Theory
Constructing Pythagorean Triples

The Pythagorean theorem states that if $\angle C$ in $\triangle ABC$ is a right angle, then

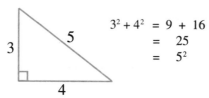

$$AC^2 + BC^2 = AB^2$$

The best-known example of a right triangle (known to the ancient Egyptians) has sides of length 3, 4, and 5.

$$3^2 + 4^2 = 9 + 16$$
$$= 25$$
$$= 5^2$$

When you first learned about the Pythagorean theorem, you probably did a lot of exercises where you came across other triplets of integers (a, b, c) that satisfy the equation $a^2 + b^2 = c^2$. A triplet of this form is known as a *Pythagorean triple*. Other well-known examples of Pythagorean triples are $(5, 12, 13)$ and $(6, 8, 10)$:

$$5^2 + 12^2 = 25 + 144$$
$$= 169$$
$$= 13^2$$

$$6^2 + 8^2 = 36 + 64$$
$$= 100$$
$$= 10^2$$

You might have noticed that $(6, 8, 10)$ is really a new version of our old friend $(3, 4, 5)$; we have just taken the 3-4-5 triangle and expanded each of its dimensions by a factor of 2. (It is misleading to say we have doubled the triangle, for although we have doubled each of the sides, we have quadrupled the area.) We can get many new Pythagorean triples in this way, by multiplying $(3, 4, 5)$ by some integer. For example,

multiplying by 100: $300^2 + 400^2 = 500^2$
multiplying by 13: $39^2 + 52^2 = 65^2$
multiplying by k: $(3k)^2 + (4k)^2 = (5k)^2$

Using $(3k, 4k, 5k)$, we can easily construct lots and lots of Pythagorean triples. Unfortunately, after the first one or two, they aren't very interesting.

128

There is in fact a method that gives *every* Pythagorean triple:

Choose three positive integers u, v, and k such that $u > v$.
Then take

$$a = k(u^2 - v^2), \quad b = 2uvk, \quad c = k(u^2 + v^2)$$

By substituting u, v, and k into the equations above, we obtain a Pythagorean triple (a, b, c).

For example, try $u = 2$ and $v = k = 1$ — you should rediscover our old friend (3,4,5). $u = 3$, $v = 2$, and $k = 1$ yields another familiar triple. We can multiply this triple by 2 to get yet another triple. Now try $u = 72$, $v = 65$, $k = 3$ and see what you get. Check that, even with such large numbers, $a^2 + b^2 = c^2$!

Food for Thought

❶. A *primitive Pythagorean triple* is an ordered triple of integers (a, b, c) that have no common factors, such that

$$a^2 + b^2 = c^2.$$

For example, (3, 4, 5), (5, 12, 13), and (65, 72, 97) are primitive Pythagorean triples, but (6, 8, 10), (15, 36, 39), and (39, 52, 65) are not.

Here are some primitive Pythagorean triples:

$(a, b, c) = $ (3, 4, 5), (5, 12, 13), (15, 8, 17), (7, 24, 25), (9, 40, 41), (21, 20, 29).

They have been arranged so that b is even.

In each of these cases, $\dfrac{c-a}{2}$, $\dfrac{c+a}{2}$, $c + b$, and $c - b$ are all perfect squares, and b is divisible by four. Try to find other primitive Pythagorean triples (perhaps using the formula). Are these facts true for your new triples as well? Do you think this is true for *all* Pythagorean triples (not just primitive ones)?

❷. Can you find all Pythagorean triples with hypotenuse $c = 60$?

❸. Check that $(u^2 - v^2, 2uv, u^2 + v^2)$ is a Pythagorean triple — in other words, that
$$(u^2 + v^2)^2 = (u^2 - v^2)^2 + (2uv)^2.$$

❹. $312^2 + 459^2 = 555^2$. Can you figure out which u, v, and k would give you this triple?

❺. If you know some trigonometry, try "breaking the rules" and substituting $u = \cos \theta$, $v = \sin \theta$, $k = 1$ in the formula on the previous page. What formula do you get? (Remember the double angle formulas: $\cos 2\theta = \cos^2 \theta - \sin^2 \theta$ and $\sin 2\theta = 2 \cos \theta \sin \theta$.)

A Strange Result in Base 2 and Base 5

Consider two lists. List A consists of the positive powers of 10 (10, 100, 1000, ...) written in base 2. List B consists of the positive powers of 10 written in base 5. Remarkably, for any integer $n > 1$, there is exactly one number in exactly one of the lists that is exactly n digits long.

Powers of 10	List A	List B
10	1010 (4 digits)	20 (2 digits)
100	1100100 (7 digits)	400 (3 digits)
1000	1111101000 (10 digits)	13000 (5 digits)
10000	10011100010000 (14 digits)	310000 (6 digits)
.	.	.
.	.	.
.	.	.

For example (setting $n = 3$), we observe there is precisely one three-digit number in the two lists — the second number in list B. And (if $n = 100$) there is just one hundred-digit number in the two lists — the thirtieth number in list A.

This result first appeared on the 1994 *Asian Pacific Mathematics Olympiad*.

A Remarkable Coincidence?
Or An Even More Remarkable Non-Coincidence?

If you try to calculate $e^{\pi\sqrt{163}}$ on your calculator, depending on how many significant digits your calculator carries, you could get the answer:

Rational Thinking

Hmmm... an irrational power of an irrational number is rational???— seems irrational to me!

262,537,412,640,768,744.00000000000

$e^{\pi\sqrt{163}}$ is not in fact an integer; if your calculator had even more significant digits, it would give its value as:

262,537,412,640,768,743.99999999999250072597...

Of course, there is no reason to suspect that a strange number like $e^{\pi\sqrt{163}}$ is actually an integer, and it seems like a remarkable coincidence that it is actually so close to an integer. What is even more remarkable is that this is no coincidence — it reflects a subtle fact in number theory.

Basically, 163 is the largest number (along with 1, 2, 3, 7, 11, 19, 43, and 67) with a special (and important) number-theoretic property (which is unfortunately to deep to delve into here). A bizarre consequence of this property is that $e^{\pi\sqrt{k}}$ is relatively close to an integer (for each k in the sequence), and the larger the k, the closer $e^{\pi\sqrt{k}}$ is to an integer. Only 43 and 67 are especially striking. (Try out other numbers on a computer. While you're at it, try $e^{\pi\sqrt{58}}$.)

In *Prime Numbers in Number Theory* (p. 124), the polynomial $n^2 - 81n + 1681$ appeared. (It takes on only prime values for $0 < n < 81$.) If you know how, take its discriminant. What do you notice? Do you think this is a coincidence? In the 1975 April Fools' issue of *Scientific American*, Martin Gardner claimed that $e^{\pi\sqrt{163}}$ is in fact an integer. This allegation caused much consternation in the mathematical community. (He also claimed that P-KR4 had been proved to be a winning move for White in Chess, that Einstein's special theory of relativity is logically flawed, and that Leonardo da Vinci had invented the valve flush toilet.)

For more surprising connections between the famous constants e and π, flip to i^i and Other Improbabilities (p. 208).

Historical Digression

The Poor Clerk Who Knew Numbers on a First-Name Basis

Srinivasa Ramanujan 1887-1920

One morning in 1913, G. H. Hardy, a distinguished fellow of Cambridge University and already a well-known mathematician, received a large untidy envelope adorned with Indian stamps. Inside the envelope were pages of rough notes containing unfamiliar looking mathematical identities — the product of a brilliantly intuitive but unschooled mind.

The originator of the letter was a poor 26-year-old clerk from Madras, India named Srinivasa Ramanujan. The manuscript was so impressive that Hardy arranged for Ramanujan to come to Cambridge to collaborate with him in mathematical research. In the following seven years, Ramanujan and Hardy established one of the most successful collaborations in the history of mathematics. Yet, in a tragic turn of events, Ramanujan contracted tuberculosis while at the peak of his creativity. As he lay on his death bed, Hardy attempted to occupy him with trivial conversation. "I drove over here today in a taxi with a boring number, 1729." Ramanujan's face brightened with the recognition of an old friend, "Oh no, that's not a boring number; it's the smallest number that's representable as a sum of two cubes in two different ways!"

Using some of the strange relationships described on the previous page, Ramanujan managed to get remarkable approximations to π, including the following:[1]

$$\pi \approx \frac{12}{\sqrt{130}} \ln\left[\frac{(2+\sqrt{5})(3+\sqrt{13})}{\sqrt{2}}\right]$$ (correct to 15 decimal places)

"ln" denotes log base *e*.

$$\pi \approx \frac{12}{\sqrt{310}} \ln\left[\frac{1}{4}(3+\sqrt{5})(2+\sqrt{2})\left[(5+2\sqrt{10})+\sqrt{61+20\sqrt{10}}\right]\right]$$ (correct to 22 decimal places)

$$\pi \approx \frac{4}{\sqrt{522}} \ln\left\{\left(\frac{5+\sqrt{29}}{\sqrt{2}}\right)^3 (5\sqrt{29}+11\sqrt{6})\left[\sqrt{\frac{9+3\sqrt{6}}{4}}+\sqrt{\frac{5+3\sqrt{6}}{4}}\right]^6\right\}$$ (correct to 31 decimal places)

---

[1]Dario Castellanos, "The Ubiquitous π", *Mathematics Magazine*. (Vol. 61, No. 2, (April 1988).), 79.

Solution to the Painted Lockers Problem (p. 127)

It turns out that all the lockers are mauve, except for the perfect squares! Here's why. Let's ignore the rest of the lockers, and see what happens to locker n throughout the day. The dth student repaints the locker if and only if d is a factor of n. (Can you see why?) So those numbers with an even number of factors are left mauve at the end of the day, and those with an odd number of factors are left vermilion. It remains only to show that the perfect squares are the only numbers with an odd number of factors. We present two arguments to show this; the first is informal while the second is a little more formal.

First argument

If n is not a perfect square, then its factors come in pairs, where the two numbers in a pair multiply to n. That is, the factor d is paired with the factor n/d, so there are an even number of factors. For example, the factors of 24 can be paired as follows: (1,24), (2,12), (3,8), (4,6). However, if n is a perfect square, $\sqrt{n}$ is paired with itself. Hence n would have an odd number of factors. For example, if n were 36, we would have the pairs (1,36), (2,18), (3,12), and (4,9), but the factor 6 would be unpaired. Hence 36 has an odd number of factors.

Second argument

We can write any number n as a product of its prime factors as follows:

$$n = p_1^{m_1} p_2^{m_2} p_3^{m_3} \ldots p_r^{m_r}$$

Each factor of n can be written in the form

$$p_1^{k_1} p_2^{k_2} p_3^{k_3} \ldots p_r^{k_r} \text{ where } 0 \le k_i \le m_i$$

Since each k_i can assume $m_i + 1$ values, n has $(m_1 + 1)(m_2 + 1) \ldots (m_r + 1)$ factors. This product is odd if and only if all the factors in the product are odd. That is, this product is odd if and only if all the values m_i are even, i.e. if and only if n is a perfect square.

In conclusion, the only lockers painted vermilion at the end of the day are those whose numbers have an odd number of factors — that is, the perfect squares.

This problem appeared on the *Twenty-Eighth William Lowell Putnam Mathematical Competition* in 1967.

Sums of kth Powers

It is a well-known fact that the sum of the first n natural numbers is given by:

$$1 + 2 + 3 + \ldots + n = \frac{n(n+1)}{2}$$

(We have already used this formula in *Two Theorems about Magic Squares*, p. 35.)
One simple proof is suggested by the following diagram:

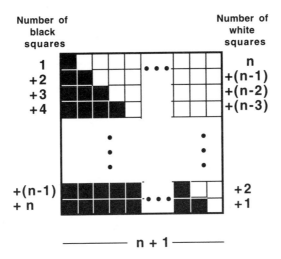

If $S_1(n)$ denotes the sum $1 + 2 + 3 + \ldots + n$, then the $n \times (n+1)$ array of squares above has $S_1(n)$ black squares and $S_1(n)$ white squares, so $2S_1(n) = n(n+1)$. Therefore,

$$S_1(n) = \frac{n(n+1)}{2}.$$

This identity $S_1(n) = \dfrac{n(n+1)}{2}$ is a special case of a more general result.

> What is the value of the following sum for a given positive integer k?
>
> $$1^k + 2^k + 3^k + \ldots + n^k$$

Let's represent this sum by $S_k(n)$. We will explore several different ways of approaching this problem in the upcoming pages.

Sums of kth Powers (cont'd)

Special case k = 2

What is the value of the sum of the first n squares, $S_2(n)$?

$$S_2(n) = 1^2 + 2^2 + 3^2 + ... + n^2$$

One of the most powerful calculating tricks for evaluating the sum of such a series is a technique called *telescoping*. This technique involves writing an expression for the difference between successive terms of a particular series and then adding all the differences together. All terms except the first and last vanish, and the sum magically appears!

Suspecting that the sum of the first n squares might be a polynomial of degree 3 in n, we write an expression for the differences between successive terms of the sequence $1^3, 2^3, 3^3, ..., n^3, ...$ starting from the last term and working backwards.

$$n^3 - (n-1)^3 = 3n^2 - 3n + 1$$
$$(n-1)^3 - (n-2)^3 = 3(n-1)^2 - 3(n-1) + 1$$
$$(n-2)^3 - (n-3)^3 = 3(n-2)^2 - 3(n-2) + 1$$

$$\begin{array}{ccccccc}
\cdot & & & & \cdot \\
\cdot & & & & \cdot \\
\cdot & & & & \cdot
\end{array}$$

$$3^3 - 2^3 = 3(3^2) - 3(3) + 1$$
$$2^3 - 1^3 = 3(2^2) - 3(2) + 1$$
$$1^3 - 0^3 = 3(1^2) - 3(1) + 1$$

Adding: $$n^3 - 0^3 = 3S_2(n) - 3S_1(n) + n$$

From which we conclude: $$S_2(n) = \frac{1}{3}(n^3 - n + 3S_1(n))$$

$$= \frac{n(n+1)(2n+1)}{6} \qquad \text{Omitting a few steps...}$$

Using the telescoping method shown above, we can derive the following formula for $S_3(n)$, where $S_3(n) = 1^3 + 2^3 + 3^3 + ... + n^3$:

$$S_3(n) = \left[\frac{n(n+1)}{2} \right]^2$$

Historical Digression

A Mathematical Genius of the Highest Order

We noted earlier that idiot savants were so-named because they demonstrated outstanding computational talents in the absence of any exceptional mathematical ability. However, the term *calculating prodigy* is a more appropriate epithet in describing Karl Friedrich Gauss, considered by historians of mathematics as one of the greatest mathematicians of all time.

Gauss enjoyed numerical computation as a child. An anecdote told of his early schooling reveals his precocious computing ability and his mathematical insight. One day, in 1787, when Gauss was only ten years old, his teacher had the students add up all the numbers from one to a hundred, with instructions that each should place his

Karl Friedrich Gauss 1777-1855

THE BETTMANN ARCHIVE

slate on a table as soon as he had completed the task. Almost immediately Gauss placed his slate on the table, saying, *"Ligget se"* ("There it lies"). The teacher looked at him scornfully while the others worked diligently. When the instructor finally looked at the results, the slate of Gauss was the only one to have the correct answer, 5050, with no further calculation. The ten-year-old boy evidently had computed mentally the sum of the arithmetic progression $1 + 2 + 3 + ...+ 99 + 100$, presumably by grouping the numbers in pairs to total 101 as shown below .

1	2	3	4	5		48	49	50
+100	+99	+98	+97	+96	$\cdots$	+53	+52	+51
101	101	101	101	101		101	101	101

There are 50 pairs, each summing to 101, for a total sum of 50×101 or 5050.

His teachers soon called Gauss's talent to the attention of the Duke of Brunswick who supported his education, first enabling him to study at the local college, then at the university in Göttingen, where he matriculated in October 1795.

Sums of kth Powers (cont'd)

The General Case: $S_k(n)$

If the sum of the kth powers of the first n positive integers is denoted by $S_k(n)$, then we observe that $S_k(n)$ is a polynomial (in n) of degree $k + 1$. (Notice in particular that $S_3(n)$ is the square of $S_1(n)$. There is no immediately obvious reason why this should be true.)

One method to calculate $S_k(n)$ given $S_{k-1}(n)$ uses telescoping series. Here is a second, more unusual method, using calculus.[1]

> •Write down $S_{k-1}(n)$ explicitly as a polynomial.
>
> •Multiply it by k.
>
> •Then take its indefinite integral with respect to n. (Make sure that you are left with no constant term; alternatively, you can integrate from 0.)
>
> •Finally, add in an appropriate linear term (in other words, a constant multiple of n) so that the sum of the co-efficients is 1.

For example, to construct $S_2(n)$ from $S_1(n)$:

$$S_1(n) = \frac{n(n + 1)}{2}$$

$$= \tfrac{1}{2}n^2 + \tfrac{1}{2}n$$

Multiply by k (in this case $k = 2$): $n^2 + n$

Take the indefinite integral: $\int (n^2 + n)\,dn = \dfrac{n^3}{3} + \dfrac{n^2}{2}$

Add the linear term $\left(1 - \dfrac{1}{3} - \dfrac{1}{2}\right)n$ to get: $\dfrac{n^3}{3} + \dfrac{n^2}{2} + \dfrac{n}{6}$.

So $S_2(n) = \dfrac{n^3}{3} + \dfrac{n^2}{2} + \dfrac{n}{6}$, which is the answer we derived earlier!

---

[1]This method was told to me in 1986 by Wayne Broughton, who was then a high school student in Vancouver, British Columbia.

Sums of kth Powers (cont'd)

Unfortunately, this technique doesn't automatically give $S_k(n)$ in a convenient factored form. For example, it's possible to prove that, when k is odd, $n+1$ is a factor of $S_k(n)$. But using this method, it is clear that $S_k(n)$ is always a polynomial of degree $k+1$ with leading coefficient $\dfrac{1}{k+1}$.

There are several other ways of calculating $S_k(n)$. One of the most unusual and effective methods uses the *Bernoulli numbers*.

(The Bernoulli numbers are a sequence of numbers denoted by B_0, B_1, B_2, ... beginning 1, -1/2, 1/6, 0, -1/30, 0, 1/42, 0, ... For more background on them, read the next section, *Bernoulli Numbers*.)

It turns out that

$$S_k(n) = \frac{1}{k+1} \sum_{j=0}^{k} \binom{k+1}{j} B_j (n+1)^{k+1-j} \quad \text{where} \quad \binom{k+1}{j} = \frac{(k+1)!}{j!(k+1-j)!}$$

and B_j is the jth Bernoulli number.

This verifies that $S_k(n)$ is a polynomial in n of degree $k+1$ as mentioned above, and you can check that the leading coefficient is indeed $\dfrac{1}{k+1}$.

Bernoulli Numbers

EXPERTS ONLY

Bernouilli numbers are a sequence of numbers that come up in a variety of circumstances. The previous section showed how they are used to find the sum of the first n k^{th} powers.

Definition. The sequence $B_0, B_1, B_2, ...,$ called the *Bernouilli numbers*, is given by $B_0 = 1$ and, for $m > 0,$

$$B_m = -\frac{1}{m+1} \sum_{j=0}^{m-1} \binom{m+1}{j} B_j \quad \text{where} \quad \binom{m+1}{j} = \frac{(m+1)!}{j!(m+1-j)!}$$

So, for example, we have:

$$B_1 = -\frac{1}{2}, \quad B_2 = \frac{1}{6}, \quad B_3 = 0, \quad B_4 = -\frac{1}{30}, \quad B_5 = 0, \quad B_6 = \frac{1}{42}$$

Bernouilli numbers display the following remarkable properties:

a) For all positive integers m, $B_{2m+1} = 0$.
B_{2m} alternates in sign for successive values of m.

b) The zeta function is defined as follows:

$$\zeta(m) = \frac{1}{1^m} + \frac{1}{2^m} + \frac{1}{3^m} + ... + \frac{1}{i^m} + ...$$

If m is a positive integer, then

$$B_{2m} = \frac{(-1)^{m+1}(2m)!\varsigma(2m)}{2^{2m-1}\pi^{2m}}$$

(proved by Euler)

> **Try This!**
> Use property b) to prove this identity.
> $$1 + \frac{1}{2^2} + \frac{1}{3^2} + ... + \frac{1}{i^2} + ... = \frac{\pi^2}{6}$$
> Can you find $1 + \frac{1}{2^4} + \frac{1}{3^4} + ... + \frac{1}{i^4} + ...$?

c) If we expand $\dfrac{x}{e^x - 1}$ as a power series in x, then

$$\frac{x}{e^x - 1} = \sum_{n=0}^{\infty} B_n \frac{x^n}{n!}$$

The Incredible Bernouilli Family

The Bernouilli family of Basel is certainly the most celebrated family in the history of mathematics. Between Nicolaus Bernouilli (1623-1708) and Jean Gustave Bernouilli (1811-1863), the family produced twelve outstanding mathematicians and physicists.

The Bernouilli numbers were discovered by Jacques Bernouilli (1654 -1705), who developed them to solve the problem of summing the kth powers of integers (see *Sums of the kth Powers*, p. 135). Bernoulli numbers are surprisingly linked to Fermat's Last Theorem (described in *The Most Famous Conjecture in Mathematics*, p. 123). The German mathematician Ernst Kummer (1810-1893) proved that if p is an odd prime that does not divide the numerators of the Bernoulli numbers B_2, B_4, ..., B_{p-3}, then Fermat's Last Theorem holds for $n = p$ (and for any n divisible by p). Such a p is called a *regular prime*. Along with the case when n is divisible by 4 (the only case that Fermat proved himself), this theorem proves Fermat's Last Theorem for all n less than 100 other than 37, 59, 67, and 74.

Jacques was also falsely credited as being the first person to notice that the following series diverges. (This series is called the *harmonic series* and is discussed more fully on pp. 179-183.)

$$1 + \frac{1}{2} + \frac{1}{3} + \frac{1}{4} + \dots$$

One of the more bizarre anecdotes about this impressive family concerns l'Hôpital's rule, which appears in virtually every elementary calculus class.

L'Hôpital's Rule:	If $f(x)$ and $g(x)$ are functions differentiable at $x = a$ such that $f(a) = g(a) = 0$, and if $\lim_{x \to a} \dfrac{f'(x)}{g'(x)}$ exists, then $\lim_{x \to a} \dfrac{f(x)}{g(x)} = \lim_{x \to a} \dfrac{f'(x)}{g'(x)}$

Jean Bernouilli (1667-1748), the younger brother of Jacques, had instructed a French marquis, G.F.A. de L'Hôpital (1661-1704), in the art of calculus. In return for a regular salary, Jean agreed to send L'Hôpital all of his mathematical discoveries. In 1694, Jean Bernouilli discovered "L'Hôpital's Rule", and the rest is history.

Nothing Makes Me Less Aware of the Passage of Time

Jordan Ellenberg (USA)
Born Oct. 30, 1971

The first mathematical experience Jordan Ellenberg remembers dates from when he was about five. After staring at a rectangular grid of holes, he suddenly realized why multiplication is commutative (that is, why it is that when you multiply two numbers together, the answer doesn't depend on the order in which you multiply them, so $m \times n = n \times m$).

When he was seven, Jordan was discovered by Eric Walstein, a teacher at the nearby Montgomery Blair High School. Jordan had already learned how to multiply three-digit numbers in his head (something he has long since forgotten), and had shown other signs of incipient talent. Walstein took Jordan under his wing and oversaw his mathematical development. In Grade 4, Jordan began participating in the *American Regions Mathematics League,* where most of the competitors were already in their final year of high school.

At about this time, Jordan discovered the following theorem:

> If a is an integer and p is a prime number, then $a^p - a$ is divisible by p.

He proved it by looking at Pascal's triangle modulo p. He soon heard, however, that his theorem had been proven before, and that it was called *Fermat's Little Theorem.* He was sorely disappointed that it was only a "little" theorem.

At Winston Churchill High School in Potomac, Maryland, Jordan really began to shine. In 1989, he placed second in the *Westinghouse Science Talent Search*, a national science fair. In his last three years of high school he did spectacularly well on the *USA Mathematical Olympiad*, rising from ninth place, to second, and finally to first. He competed for the United States at the *International Mathematical Olympiad* in each of those three years, winning two Gold Medals and one Silver Medal.

While studying mathematics as an undergraduate at Harvard, Jordan competed in the North American *Putnam Mathematical Competition*, twice achieving the highest honor of Putnam Fellow. In three of his four years, he was a member of the Harvard team that was in the process of winning the Putnam competition for eight consecutive years.

Despite his mathematical successes, Jordan spent much of his undergraduate career avidly pursuing his interest in literature. He regularly wrote for the *Harvard Crimson* and was the Fiction Editor for the *Harvard Advocate*. After graduation, he completed a Masters Degree in Creative Writing at Johns Hopkins University in Baltimore, where he wrote a novel, *The Grasshopper King*. When asked about this unusual detour, Jordan responds, "It is always good to have some element of resistance to what's easy and what's obvious." He always intended to return to mathematics, and this was his last opportunity to write seriously. After completing his Masters, he returned to Harvard to begin a Ph.D. in Number Theory.

Jordan has strong views on the role of mathematics in society: "Math occupies a strange cultural place. If you have an aptitude for math, that often makes you the 'brain' of your class, more so than for other fields such as history or writing." He believes that mathematics is too often seen as the province of a special class of people. There is even a disturbing cheerfulness to people's admission that they don't like math, and aren't good at it. As Jordan reflects, "You don't hear people say, 'Long words make my head spin!'"

But everyone needs basic math skills to be a functioning member of society. As a result, Jordan feels that the strength of a mathematical curriculum should not be judged on the basis of abstruse theorems in calculus. People should learn to use mathematical thinking to make them better critics of the world around them. Jordan cites newspaper polls as one example: "In order to tell when you are being manipulated, you need to have some feel for probabilities and quantities. What does it mean when you hear that '50% of Bud drinkers would prefer Miller'? What does it mean when you hear that the 'rate of growth of the deficit is decreasing'?"

For Jordan, the most remarkable feature of mathematics is its ability to completely engage his mind and soul: "Hours can go by. Nothing makes me less aware of the passage of time." His parting advice for young people interested in the subject? "Do math for math's sake, not because your parents will be proud of you, or because people will think you are smart."

FIBONACCI
&
THE GOLDEN MEAN

The merit of painting lies in the exactness of reproduction. Painting is a science and all sciences are based on mathematics. No human inquiry can be a science unless it pursues its path through mathematical exposition and demonstration.

Leonardo da Vinci

Fibonacci & The Golden Mean

Funny Fibonacci Facts

For the definition of the Fibonacci numbers see Fibonacci: The Greatest European Mathematician of the Middle Ages (p. 57).

Here are some unusual facts about the Fibonacci numbers. If you read *A Formula for the nth Term of the Fibonacci Sequence* (p. 60) or *The Golden Mean* (p. 150), you will see some of them proved, and you might get ideas about how to prove others.

1. As n gets larger, the number $\dfrac{F_{n+1}}{F_n}$ gets closer to $\tau = \dfrac{1+\sqrt{5}}{2}$, the golden mean.

2. a) $F_n^2 + F_{n+1}^2 = F_{2n+1}$ for $n \geq 1$.

   b) $F_{n+1}F_{n-1} - F_n^2 = (-1)^n$ for $n \geq 2$.

   c) $F_{n+1}^3 + F_n^3 - F_{n-1}^3 = F_{3n}$ for $n \geq 2$.

   d) If m is a factor of n, then F_m is a factor of F_n.

   e) Let $G_n = \dfrac{F_{2n}}{F_n}$. (It follows from part d) that G_n is an integer.)

   Then $G_n = G_{n-1} + G_{n-2}$.

   f) $\displaystyle\sum_{n=2}^{\infty} \arctan\left(\dfrac{(-1)^n}{F_{2n}}\right) = \dfrac{1}{2}\arctan\dfrac{1}{2}$.

3. Bees have unusual reproductive habits. Each male bee has only one parent (a mother, the queen bee), whereas each female bee has two parents (both a mother and a father). So each male bee has one parent, and you can quickly check that he has two grandparents (the two parents of his mother). How many great-grandparents does he have? How many great-great-grandparents? Can you guess the pattern? Can you prove it? (Drawing a family tree might help.)

4. a) Can you prove this identity, $F_1^2 + F_2^2 + \ldots + F_n^2 = F_n F_{n+1}$?

   This diagram may suggest a proof to you.

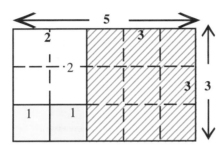

   b) Can you find a simple formula for $F_1 + F_2 + \ldots + F_n$ in terms of F_{n+2}?

   It might remind you of the formula for $2^0 + 2^1 + \ldots + 2^n$ in terms of 2^{n+1}.

Try This!

5. Use a calculator to evaluate 10000/9899 to eight or ten decimal places. What pattern do you notice? Can you guess the next few digits? (The remaining questions will make no sense unless you have already guessed the pattern.) The digits must eventually repeat, as 10000/9899 is a rational number. Do you find this surprising? Can you find more numbers with a similar property? (Hint: Try 100/89. Then try to find more.) What if you wanted powers of 2 to appear instead?

6. Consider a circle whose circumference is the *golden mean*, $\tau = \dfrac{1+\sqrt{5}}{2} \approx 1.61803$.

Start at any point, on the circle and take some number of consecutive steps of arc length one in the clockwise direction. Number the points you step on in the order you encounter them, labelling your first step P_1, your second step P_2, and so on. Prove that when you stop, the difference in the subscripts of any two adjacent numbers is a Fibonacci number.

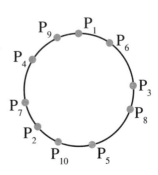

7. The remarkable interplay between Fibonacci numbers and Pascal's triangle is best shown in a diagram without any words of explanation:

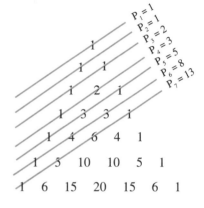

There are several ways of proving why this is true. What may be the easiest method is to let P_n denote the nth *diagonal Pascal sum*, and then observe that $P_1 = P_2 = 1$ and show that $P_n = P_{n-1} + P_{n-2}$ for $n > 2$ (which is not obvious).

Another method is to use the "Elvis" way of looking at Fibonacci numbers. (See any of the sections dealing with *Elvis Numbers.*) In how many ways can Elvis get to the nth step? Well, we know (from *The Solution to Elvis the Elf's Eccentric Exercise,* p. 59) that he can manage this in F_n ways. But let's try an alternate way of looking at it. He can take n single steps and 0 double steps (which he can do in $\binom{n}{0} = 1$ way); or he can take n-2 single steps and 1 double step (which he can do in $\binom{n-1}{1} = n - 1$ ways, because out of the n-1 steps he takes, he must choose one to be the double step); or he can take n-4 single steps and 2 double steps (which he can do in $\binom{n-2}{2}$ ways, because out of the n-2 steps, he must choose two of them to be double steps); and so on. The total number of ways in which he can get to the nth step is thus

$$\binom{n}{0} + \binom{n-1}{1} + \binom{n-2}{2} + \ldots$$

which is just P_n. (Just look at the diagram again if you don't believe it!) Thus P_n is the nth Elvis number, which is F_n!

8. Elvis the Elf decided to create a landing of marble tiles at the foot of his front hall staircase. He measured off a rectangular area of length 13 feet and width 5 feet and determined that he would need 65 marble tiles measuring one foot square to cover this landing. He purchased 65 tiles but upon arriving home discovered that one tile was damaged beyond repair and only 64 of the tiles could be used. Undaunted by the challenge, he arranged the 64 tiles in an 8 × 8 square array and then cut the large square into four regions, A, B, C and D as shown below left. He then rearranged these four pieces into the 13 × 5 rectangle shown below right. *How did Elvis the Elf create a rectangle of area 65 square feet by rearranging the pieces of a square of area 64 square feet?*

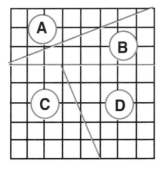

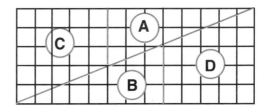

Take a piece of graph paper, and cut out an 8 × 8 square. Then cut the square into four pieces as shown above. Rearrange these four pieces to form a 5 × 13 rectangle. What do you discover?

Notice that the numbers {5, 8, 13} are consecutive Fibonacci numbers. Try to create a paradox like the one above by replacing them with another triplet of larger consecutive Fibonacci numbers, such as {8, 13, 21}. You might notice a similarity to fact # 2 a).

I learned of some of the more unusual results in fact # 2 from Georg Gunther of Memorial University of Newfoundland. The bee problem is from H. R. Jacobs, *Mathematics, A Human Endeavor.* p. 98-99. Fact # 6 was related to me by Greg Kuperberg of the University of Chicago. Fact # 8 is taken from H. S. M. Coxeter, "The Golden Section, Phyllotaxis, and Wythoff's Game", *Scripta Mathematica.* (Vol. 19, No. 2-3, (June-Sept. 1953).), 135-43.

The Golden Mean

The *golden mean* is the number $\tau = \dfrac{1+\sqrt{5}}{2}$ ($\approx$ 1.618033989), often represented by the greek letter τ (spelled "tau", and pronounced so it rhymes with "Yow!"). The ancient Greeks felt that the $1 \times \tau$ rectangle was the most aesthetically pleasing, and much classical architecture is based on that proportion. (Any rectangle similar to this one is called a "golden rectangle".)

THE BETTMANN ARCHIVE

The Parthenon of Ancient Greece, completed in 438 B.C., was constructed so that its front elevation forms a golden rectangle. But nature also uses the golden mean in its own architecture.[1]

$\leftarrow$ The aesthetically pleasing $1 \times \tau$ rectangle

τ, sometimes called the *golden section* or *golden ratio*, is the first letter of the ancient greek word τομή meaning "the section". τ has already appeared in our discussions about the Fibonacci sequence, where we saw that:

$$F_n = \frac{\tau^n - (1-\tau)^n}{\sqrt{5}}$$

---

[1]In Coxeter's "The Golden Section, Phyllotaxis, and Wythoff's Game", further unusual occurrences of the golden mean are discussed, including a surprising appearance of Fibonacci numbers in pineapples. (For bibliographic information, see the *Annotated References*.)

Some Interesting Properties of the Golden Mean

The formula for Fibonacci numbers (p. 60) is just one of a series of surprising situations in which τ mysteriously appears. Here are some of them, many of which you can try to prove yourself.

1. a) τ is an irrational number; in other words, there are no positive integers m and n such that $\tau = m/n$. But τ can be approximated by rational numbers. For example, $3/2 = 1.5$ is pretty close; $55/34$ ($\approx 1.61764...$) is closer still. In fact, the "best" rational approximations to the golden mean are given by ratios of consecutive Fibonacci numbers!

   b) A related fact is that the fraction $\dfrac{F_{n+1}}{F_n}$ converges to τ as n gets large.

   c) In fact, if you take any two positive real numbers a and b, and define the sequence G_n by $G_1 = a$, $G_2 = b$, and $G_n = G_{n-1} + G_{n-2}$, then the fraction $\dfrac{G_{n+1}}{G_n}$ gets closer and closer to τ as n gets large.

2. a) $\tau^2 - \tau - 1 = 0$.

   b) Using this equation, you can show that if you take a golden rectangle ABCD (where AB is the short side), and subtract a square, ABEF, then the remaining rectangle FECD is also golden.

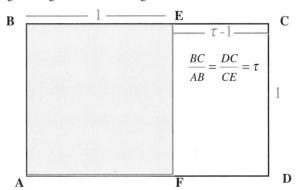

This fact is related to fact # 4 a) in *Funny Fibonacci Facts* (p. 146).

Some Interesting Properties of the Golden Mean (cont'd)

3. Here are a couple of calculator experiments (that can also be done on a computer).

a) Choose any positive number. Write it down on a piece of paper, and enter it on your calculator. Then press the following keys:

Record the new number. Repeat by pressing the same sequence of keys, and then once again write down the number you get. Keep doing this. What do you notice?

b) Repeat the process in part a), but press this sequence of keys:[1]

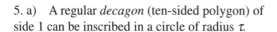

4.a) $\left(\dfrac{1}{1}-\dfrac{1}{2}-\dfrac{1}{3}+\dfrac{1}{4}\right)+\left(\dfrac{1}{6}-\dfrac{1}{7}-\dfrac{1}{8}+\dfrac{1}{9}\right)+\left(\dfrac{1}{11}-\dfrac{1}{12}-\dfrac{1}{13}+\dfrac{1}{14}\right)+... \quad =\dfrac{2}{\sqrt{\tau}}\log\tau$

b) $\sin 18° = \dfrac{\tau-1}{2}$

c) $\cos 36° = \dfrac{\tau}{2}$

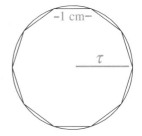

5. a) A regular *decagon* (ten-sided polygon) of side 1 can be inscribed in a circle of radius τ.

b) The icosahedron is a platonic solid with twenty faces. It will be familiar to some readers as a twenty-sided die. If A and B are two "neighboring" vertices, and A and C are two vertices "once removed", then

$$\dfrac{AC}{AB} = \tau.$$

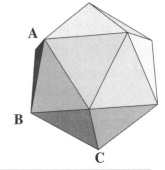

---

[1]This sequence is not universal to all calculators.

Some Interesting Properties of the Golden Mean (cont'd)

6. By pure coincidence (or some evil conspiracy), 1 mile is approximately equal to τ kilometres. This means that you can use Fibonacci numbers to quickly translate between miles and kilometres. For example, 13 km is about 8 miles; 89 km/h is approximately 55 mph.

7. (More experimentation.)

a) The *greatest-integer function* [ ] is defined as follows:
[x] is the integral part of x. For example, [9.7] = 9, and [π] = 3.
Make two lists of integers:

$$[\tau], \ [2\tau], [3\tau], \ \ldots$$

$$[\tau^2], [2\tau^2], [3\tau^2], \ \ldots$$

What do you notice?

b) (You will only need about four decimal places of accuracy for this one.)
Make another list of real numbers:

$$\tau, \ \tau^2, \ \tau^3, \ \tau^4, \ldots$$

What do you notice?

c) Do the same thing as in part b) with the sequence:

$$\tau/\sqrt{5}, \ \tau^2/\sqrt{5}, \ \tau^3/\sqrt{5}, \ \tau^4/\sqrt{5}, \ldots$$

A strange fact about these strange properties: a surprising number of them have in some way to do with the number 5. 5 appears in the definition of τ. It also appears in properties #4 (notice that $18° = 90°/5$), #5, and #7 c).

Many of these properties are discussed elsewhere (see *Funny Fibonacci Facts*, p. 146, and *The Pythagorean Pentagram, The Golden Mean, and Strange Trigonometry*, p.154). If you want to learn more about the golden mean, you should visit these pages. But if you haven't already done so, you should first check out the very first section, *Elvis Numbers* (p. 56).

153

The Pythagorean Pentagram, The Golden Mean, & Strange Trigonometry

The golden mean τ $(\tau = \dfrac{1 + \sqrt{5}}{2} \approx 1.618)$ is one of those mathematical constants (such as π or e) that pops up in the most unusual places, as we have seen in the previous section. For example, if you use your calculator to compute $2\cos 36°$ or $2\sin 18° + 1$, you'll get τ. If you compute the reciprocal of $2\sin 18°$, you'll get τ again. Let's see why this is true.

Consider the regular pentagon with all of the diagonals drawn in, and labelled as follows:

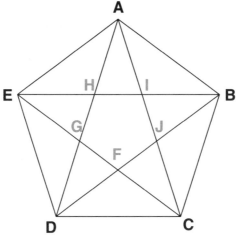

This figure, called a *pentagram*, was the symbol of the Pythagorean brotherhood. The Pythagoreans were a secret society dedicated to the pursuit of mathematics and philosophy, established by Pythagoras of Samos (c. 580-500 B.C.) on the southeastern coast of what is now Italy.

With a little thought, you can fill in all the angles. (Hint: start by showing that the angle between consecutive sides of a pentagon is 108°.)

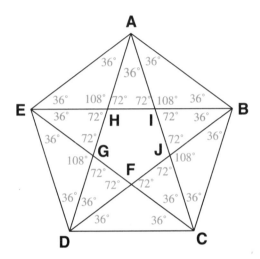

We choose our unit of length so that the pentagon has side of length 1. Therefore, AB = 1. Let t denote the length of AC.

By checking angles, you'll see that $\triangle AIB$ is similar to $\triangle ABC$, so $\dfrac{AI}{AB} = \dfrac{AB}{AC}$.

Since AB=1 and AC= t, $AI = \dfrac{1}{t}$.

$\triangle IBC$ is isosceles, so IC = BC. BC is 1, so IC = 1.

$t = AC = AI + IC = \dfrac{1}{t} + 1$, so $t^2 - t - 1 = 0$.

The roots of this quadratic equation in t are $\dfrac{1 + \sqrt{5}}{2}$ and $\dfrac{1 - \sqrt{5}}{2}$.

As t is a length, it is positive, so we can throw out the second solution. Therefore,

$t = \dfrac{1 + \sqrt{5}}{2}$, the golden mean!

Equipped with this information, we have two basic triangles: ΔACD and ΔABC.

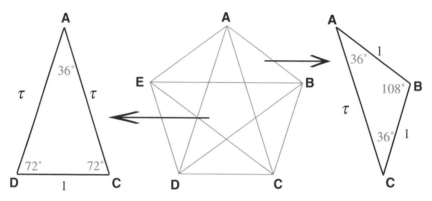

You should be able to see lots of triangles similar to both of these in the pentagram. For example, ΔEAB and ΔABC are equiangular and therefore similar. From here, we can work out lots of trigonometric values. For example, if we drop an altitude from A to EB in ΔEAB, as shown below,

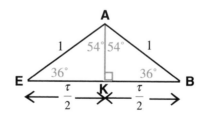

we see that $\cos 36° = \dfrac{EK}{EA} = \dfrac{\tau}{2}$

You can derive a lot more yourself. The next page provides some ideas to get you started.

Food for Thought

❶. Apply the cosine law to one of the basic triangles to compute cos 108°.

Apply the sine law to compute $\dfrac{\sin 72°}{\sin 36°}$, and (using sin 72° = 2 sin 36°cos 36°) compute cos 36° in another way.

❷. Prove that $\sin 18° = \dfrac{\tau - 1}{2}$. Compute cos 72° and sin 54°.

❸. For the original pentagram, show that $\dfrac{AC}{AB} = \dfrac{AB}{AI} = \dfrac{AI}{IJ} = \tau$.

Compute the ratio $\dfrac{\text{Area ABCDE}}{\text{Area FGHIJ}}$.

❹. A regular decagon (10-gon) of side 1 is inscribed in a circle. What is the radius of the circle? (Hint: See *Some Interesting Properties of the Golden Mean*, p. 151.)

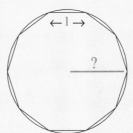

❺. Using the relationships in exercise 3, and the fact that the ratios of consecutive Fibonacci numbers (especially large ones) are approximately τ, we can fill in lengths of the pentagram that are approximately integers, and Fibonacci numbers to boot. Can you determine the missing length? (Warning: these are only approximations!)

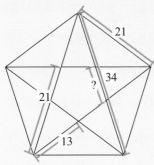

For an easy way to compute other odd trigonometric values, such as

$$\sin 15° = \frac{\sqrt{3}-1}{2\sqrt{2}},$$

*flip to **The Ailles Rectangle** (p. 87).*

157

GEOMETRY REVISITED

That vast book which stands forever open before our eyes; I mean the universe, cannot be read until we have learned the language. It is written in mathematical language, and its characteristics are triangles, circles and other geometric figures, without which it is humanly impossible to comprehend a single word; without these one is wandering about a dark labyrinth.

— Galileo Galilei

A "Do-It-Yourself" Proof of Heron's Formula

Heron's formula provides a quick way of computing the area of a triangle given the lengths of the sides. Let the sides of the triangle be denoted by a, b, and c, and let s denote the semi-perimeter, where s is defined by:

Heron's Formula

$$s = \frac{a+b+c}{2}$$

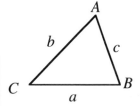

If the area of the triangle is K, then

$$K = \sqrt{s(s-a)(s-b)(s-c)}$$

Heron (c. 75 A.D.) was a mathematician from Alexandria (in modern-day Egypt). Although the formula bearing Heron's name was known to Archimedes several centuries earlier, the proof by Heron is the earliest known to us.

In many parts of North America, this formula is still taught in high school, but very few people know why it is true. Here is a do-it-yourself proof; the key steps are indicated, but you'll have to fill in the blanks yourself.

Step 1

We'll need a little notation first. Let ABC be the triangle in question, with sides a, b, c. The values s-a, s-b, and s-c will come up repeatedly, so we'll define them as new variables. Let $x = s$-a, $y = s$-b, $z = s$-c.

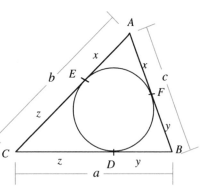

Show that $a = y + z$, $b = x + z$, $c = x + y$, and $s = x + y + z$.

Let D, E, and F be the points of tangency of the inscribed circle with sides BC, CA, and AB respectively.
Show that $AE = AF = x$, $BF = BD = y$, and $CD = CE = z$.

Step 2

Next, we'll get two formulas for the area K. Let r be the radius of the inscribed circle. Show that $K = rs$ as follows. Let I be the incenter. Work out the areas of IAB, IBC, and ICA in terms of r, a, b, c, and add the results together to get K.

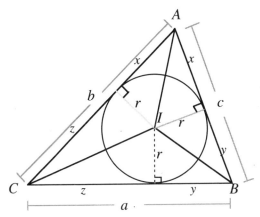

Let r_a be the radius of the escribed circle opposite A. (The escribed circle away from A is defined to be the circle tangent to all three sides of ABC, lying outside the triangle, away from A, as shown below.) Let I_a be its center.

Show that $K = r_a x$ by working out the areas of I_aAB, I_aBC, and I_aCA and showing that if $K = \text{Area}(ABC)$ then:

$$K = \text{Area}(I_aBA) + \text{Area}(I_aCA) - \text{Area}(I_aBC)$$

Similar arguments show that $K = r_b\, y = r_c\, z$ where r_b and r_c are the radii of the escribed circles opposite B and C respectively.

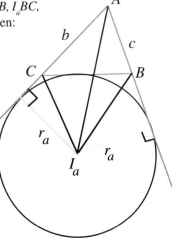

161

Step 3

Let G be the point of tangency of the escribed circle opposite A to BC, as in the figure below. Show that $BG = z$. (This might be tricky, depending on how you go about it.)

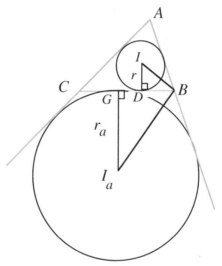

Show that triangle BGI_a is similar to triangle IDB. Conclude that $\dfrac{ID}{DB} = \dfrac{BG}{GI_a}$, and therefore that $rr_a = yz$.
Similar arguments show that $rr_b = zx$ and $rr_c = xy$.

Step 4

Using Step 2, show that $K^6 = (rs)\,(rs)\,(rs)\,(r_a\,x)\,(r_b\,y)\,(r_c\,z)$

By using the identities from Step 3 ($rr_a = yz$, $rr_b = zx$, $rr_c = xy$) and a little algebra, show that:
$$K^6 = (sxyz)^3$$

From there, recalling that $x = s - a,\ y = s - b,\ z = s - c$, prove Heron's formula:

$$K = \sqrt{s(s-a)(s-b)(s-c)}$$

Congratulations — you've done it!

A Short Route to the Cosine Law

If you know (and understand) the dot product, you can prove the Cosine Law really quickly. (This is a handy way of remembering the Cosine Law!)

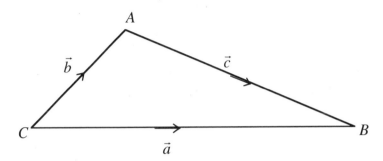

Theorem (Cosine Law)

In any $\triangle ABC$:
$$a^2 = b^2 + c^2 - 2\,bc\,\cos A$$
$$b^2 = a^2 + c^2 - 2\,ac\,\cos B$$
$$c^2 = a^2 + b^2 - 2\,ab\,\cos C$$

where a, b, and c are the lengths of the sides opposite angles A, B, and C respectively.

Proof.   We prove the last equation; the proofs of the others are the same.

$$c^2 = \vec{c} \bullet \vec{c}$$
$$= (\vec{a} - \vec{b}) \bullet (\vec{a} - \vec{b})$$
$$= \vec{a} \bullet \vec{a} + \vec{b} \bullet \vec{b} - 2\,\vec{a} \bullet \vec{b}$$
$$= \vec{a} \bullet \vec{a} + \vec{b} \bullet \vec{b} - 2\,ab\cos C$$
$$= a^2 + b^2 - 2\,ab\cos C$$

If you have only seen the properties of the dot product proved using the Cosine Law, this isn't technically a proof, of course. But it shows how integrally the two concepts are related.

163

Solution to the Falling Ladder Problem (p. 85)

Surprisingly enough, the flashlight describes a quarter circle.
Here's why. Picture the ladder at some point in mid-fall.

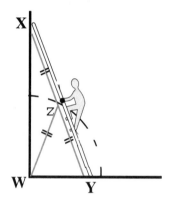

Label the points in the diagram W, X, Y, Z as shown, where W is where the wall meets the ground, X is where the ladder meets the wall, Y is where the ladder meets the ground, and Z is the location of the flashlight. $XZ = YZ$, since the flashlight is is at the midpoint of the ladder. Let this distance be l.

Using simple Euclidean arguments, it's not difficult to prove that $WZ =$ l as well. So, as the ladder falls to the ground, the flashlight is always a distance l from point W. This seems very much like the definition of a circle!

More generally, if the burglar weren't exactly halfway up the ladder, the flashlight would travel through a quarter ellipse. This problem is an example of a "locus problem". The locus of a point is the path of a point as something happens. In this problem, we're wondering about the locus of the midpoint of the ladder as the ladder falls to the ground.

You can read about other (seemingly unrelated) locus problems in the next section.

I am thankful to Ed Barbeau of the University of Toronto for introducing me to this classic.

Locus Hokus Pokus

In The Falling Ladder Problem (p. 85), we briefly introduced the concept of "locus". Here are two locus-related problems with surprising answers to test your mettle.

❶. In a circle C of radius r ($r > 10$), a chord of length 10 travels around the perimeter of the circle. The midpoint of the chord traces out another circle D. What is the area lying between C and D?

❷. Fix a circle C in the plane. Take another circle B with half the diameter of C, and "roll it around the inside" of circle C. Fix a point $\mathcal{P}$ on B. How does $\mathcal{P}$ move as B rolls around C?

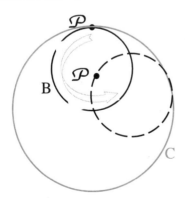

The surprising answers to these problems — without proofs — appear in More Locus Hokus Pokus (p. 173).

The Triangle Inequality
From "Common Sense" to Einstein

If you have begun to learn calculus, you have probably struggled with ε–δ proofs until your head started to spin. If so, you have some insight into the importance of inequalities in mathematics, and you won't need to be convinced that inequalities are often as difficult and ugly to prove as they are useful.

However, some inequality results are as beautiful as they are simple to state. Perhaps the simplest and most beautiful is the Triangle Inequality.

The Triangle Inequality, First Version

The shortest distance between two points in a plane is a straight line.

This may seem like common sense to you. Furthermore, the title *Triangle Inequality* might seem perplexing as no triangle appears in the statement of the theorem. The following rephrasing may help.

The Triangle Inequality Rephrased

Let P, Q, and R denote any three points. Then |PQ| $\leq$ |PR| + |RQ|.
Equality holds if and only if R lies on the line segment joining P and Q.

Proof of the Rephrased Triangle Inequality Using the Original Version

The shortest path from P to Q is the line segment joining them. Thus the path from PQ consisting of the path from P to R, followed by the path from R to Q, must be at least as long. That is, |PQ| $\leq$ |PR| + |RQ|.

Furthermore, if equality holds, then the path through R must be the shortest path. In other words, the path through R must be the straight-line path from P to Q, so R must have been on the line segment joining P and Q in the first place.

We can use the triangle inequality to solve the following real-life problem.

Farmer Brown stands 2 km away from a (straight) river. Einstein, his prize cow, is standing 1 km from the river, 6 km downstream from Farmer Brown. Brown wants to go to the river to get some water and bring it to Einstein. What is the shortest route he could take?

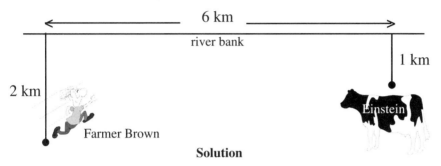

Solution

Clearly he should make a beeline for some point on the river bank, and then head back directly to Einstein. The crucial question is: which point on the river bank?

We will see that the he should head for the point 4 km along the river bank from where he's standing. Label the points as shown in the diagram below. Let E' denote any point on the river.

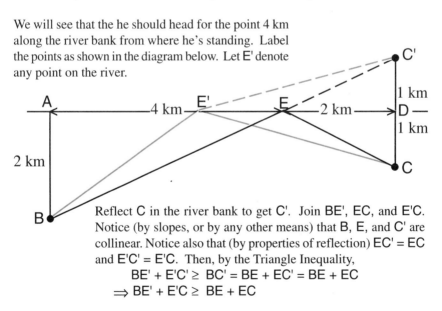

Reflect C in the river bank to get C'. Join BE', EC, and E'C. Notice (by slopes, or by any other means) that B, E, and C' are collinear. Notice also that (by properties of reflection) EC' = EC and E'C' = E'C. Then, by the Triangle Inequality,

$$BE' + E'C' \geq BC' = BE + EC' = BE + EC$$
$$\Rightarrow BE' + E'C \geq BE + EC$$

The left side is the distance that Farmer Brown would have to travel if he aimed for E'. The right side is the distance he would have to travel if he aimed for E. This proves that the route by way of E is at least as good as any other route. Q.E.D.

167

You might have noticed in the previous problem that ΔBAE is similar to ΔCDE, so ∠AEB = ∠DEC. Thus if there were a ball at B and a wall at AD, and if you pushed the ball to bounce off the wall at E, then it would follow the path from E to C. This reflection principle works in general. That is, when a ball bounces off a wall, the angle of reflection (∠DEC) is equal to the angle of incidence (∠AEB). Furthermore, the path BEC is the shortest of all paths from B to C by way of the wall.

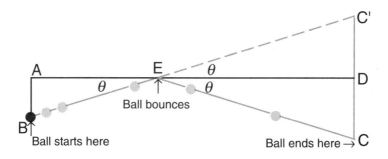

Something similar happens even when the wall is curved, which leads to some unusual results. For example, imagine that you had a pool table in the shape of an ellipse.

The equation of an ellipse can be given by:

$$\frac{x^2}{a^2} + \frac{y^2}{b^2} = 1$$

where *a* and *b* are non-negative real numbers and *a* > *b*.

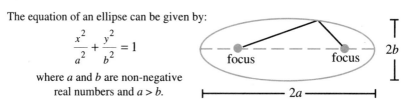

There are two points in the ellipse called *foci* (the plural of *focus*) situated symmetrically about the center, along the longer axis.

If the ball is at one focus and there is a hole at the other, then no matter which direction you hit the ball, it will bounce off a wall and go in the hole. You can't lose! (What if the ball isn't at the focus? In what direction should you hit it so that it banks off a wall and goes in the hole? What happens if the table is shaped like a parabola?)

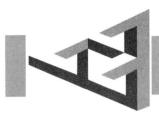
Sound waves in an elliptically shaped room display the same reflection characteristics as billiard balls on an elliptical pool table. Any sound emitted at one focus of the elliptical room can be heard with remarkable clarity at the other focus. The whispering gallery in the Tower of London has two foci a considerable distance apart and yet a whisper at either focus can easily be heard at the other focus. John Quincy Adams (1767-1848), the sixth president of the United States, understood this principle and placed his desk at one focus of the elliptical *Old House Chamber* in the Capitol building in Washington D. C. He placed the meeting table at the other focus. In this way he was able to eavesdrop on the conversations of his colleagues and monitor their loyalties.

All this talk of reflection may also remind you of how light behaves. As we will soon see, this analogy can be pursued much farther. Imagine now that Farmer Brown is standing in a field of grass, and he sees Einstein standing in a field of aromatic manure in an adjacent field. Once again, Farmer Brown wants to get to Einstein as quickly as possible. Naturally, he can move much faster over grass than he can through manure.

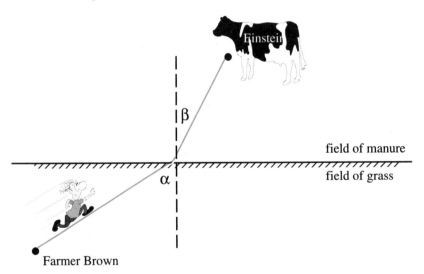

He is best off making a beeline to some point on the boundary between the grass and the manure. Because he moves faster over grass, he will choose to spend a greater proportion of time in the grass than he would have had he headed straight for Einstein.

169

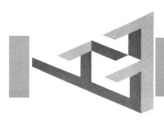

If you've seen *Snell's Law* in optics, you will recognize Farmer Brown's situation. If light moves at different speeds in two different media, it will "bend" (or refract) at the boundary according to a simple rule. In fact, in this situation there is a similar rule ("Smell's Law?") relating the angles α and β to Farmer Brown's speeds over grass and manure.

Snell's Law

> When a beam of light passes from medium 1 at an angle α to a perpendicular to the interface, it enters medium 2 at an angle β, where α and β are related by the equation,
>
> $$\frac{\sin \alpha}{\sin \beta} = \frac{v_1}{v_2}$$
>
> and v_1 and v_2 denote the velocities of light in medium 1 and medium 2 respectively.

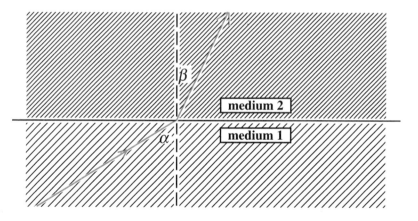

What if, instead, there were no distinct boundary between the grass and the manure? What if the manure just got deeper and thicker? In this case, we can well imagine that Farmer Brown's optimal path would be curved, as in the figure.

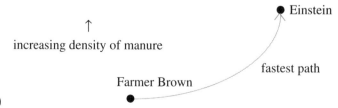

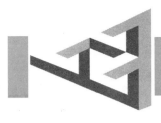

Geometry Revisited

How does all this discussion tie into our original question, "What is the shortest distance between two points?" What happened to our common sense proposition that "the shortest path between two points is a straight line"? Must we discard it? It may surprise you to learn that the answer is *no*—we just change the definition of straight line to mean *quickest path*. (The technical name for this is *geodesic*.) Then we can transfer our notions of straight lines from ordinary flat two-dimensional space to odder spaces such as the surface of a sphere (where geodesics are *great circles*) and Farmer Brown's field of manure.

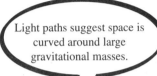

Light paths suggest space is curved around large gravitational masses.

The theory of *General Relativity* developed by Albert Einstein (not the cow!) implies that the space we live in isn't flat. Light is attracted by gravity and follows a path that curves toward gravitational sources. Before Einstein, this fact seemed to contradict the belief that light takes the shortest path and still travels in straight lines. Einstein's theory asserts that light indeed travels in geodesics, but that space becomes curved (much less "flat") near gravitational sources. That is, the shape of the path traveled by a ray of light is determined by the geometry of the space in which it is traveling.

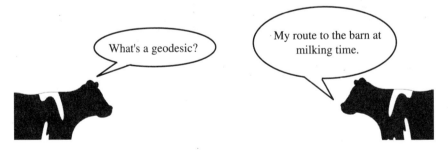

What's a geodesic?

My route to the barn at milking time.

This field of mathematics is very active today. It falls into many different categories — differential geometry and mathematical physics among others — and research on these issues will continue to shape the way we view our universe.

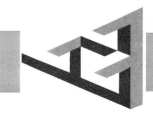

Food for Thought

❶. The U.S. Chess Federation is planning a tournament for all the Grandmasters in the country. It wants to locate the tournament so the total distance traveled by all the players is a minimum. More than half of the Grandmasters live in New York City. Can you *prove* that the best site to hold the tournament is in New York regardless of the locations of the other players? (Assume that New York City is a point, that the U.S. is flat, and that the players always take direct routes and never get caught in New York traffic.)

❷. a) Choose some general point P on side AB of an equilateral triangle ABC. How would you construct the shortest path P → Q→ R → P so that Q lies on side BC and R lies on side AC?

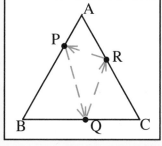

b) (This is quite tricky!) Given an acute triangle ABC, we construct a triangle PQR with P on AB, Q on BC, and R on CA so that ΔPQR has minimal perimeter. Show that P is the foot of the altitude from C to AB, Q is the foot of the altitude from A to BC, and R is the foot of the altitude from B to CA.

More Locus Hokus Pokus

In *Locus Hokus Pokus* (p. 165), two problems were posed. The answers might surprise you.

❶. In a circle C of radius r ($r > 10$), a chord of length 10 travels around the perimeter of the circle. The midpoint of the chord traces out another circle D. What is the area lying between C and D?

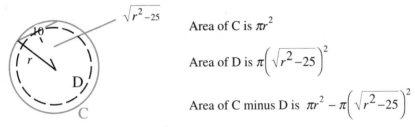

Area of C is πr^2

Area of D is $\pi \left(\sqrt{r^2 - 25} \right)^2$

Area of C minus D is $\pi r^2 - \pi \left(\sqrt{r^2 - 25} \right)^2$

The answer is 25π. Surprisingly, the answer is independent of r!

Here is a famous problem that is similar. A length of string is wrapped tightly about the earth's equator. Then the string is cut, a length of 1 meter is added, and the string is retied. It is then suspended a fixed distance above the earth's surface. What is that distance? (Assume the earth is a perfect sphere.)

❷. Fix a circle C in the plane. Take another circle B with half the diameter of C, and roll it around C. Fix a point $\mathcal{P}$ on B. How does $\mathcal{P}$ move as B rolls around C?

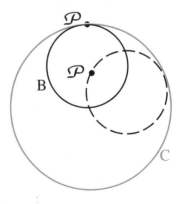

The point $\mathcal{P}$ travels along a fixed diameter of C. All of the circular motion is somehow transformed into linear motion! (This unusual result might not come as a surprise to those readers who have played with *Spirograph*™ or worked with gears.)

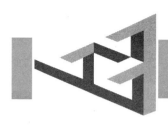
There is an even more surprising generalization to the first problem. Take a large random convex shape C, large enough so that a chord of length 10 can travel around the perimeter. The midpoint of the chord will sweep out another odd shape D. (The convex condition, which essentially means that C doesn't have any indentations, is there to ensure that D actually sits inside C.) It turns out that the area between C and D is always 25π, regardless of the shape of C!

One proof of this general result uses a clever application of Green's Theorem, often taught in second-year university calculus. But with much less machinery, it is possible to prove special cases of this result. The first problem from *Locus Hokus Pokus* proves the theorem when C is a circle. If you try to prove it when C is a rectangle, you will discover *The Falling Ladder Problem* (p. 85) in a new guise. How about other shapes? An equilateral triangle?

A Short Question on Symmetry

This is an old classic: Why is it, when you look into a mirror, that left and right are reversed but top and bottom are not? (If you think you know, try explaining it to a friend.)

INFINITY

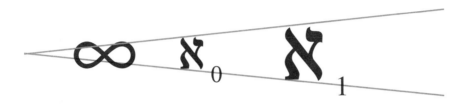

*The infinite! No other question has ever
moved so profoundly the spirit of man.*

— David Hilbert

Infinity

The Mathematics of the Birds and the Bee

Two birds are racing towards each other in the heat of passion. They are initially 10 km from each other, and they are flying at the speed of 0.5 km/min. A voyeuristic bee, which can fly at a speed of 1 km/min, starts with one of the birds and flies towards the other. When it reaches the second bird, it turns around and flies back towards the first bird again. The bee keeps this up until the two birds meet.

How far does the bee travel before this happens? (The answer is on the bottom of page 184.)

Infinity

An Early Encounter with the Infinite

The paradoxes of Zeno of Elea, a philosopher in ancient Greece, reveal some early encounters with the notion of infinity and its strange properties. About 450 B.C., he made the following assertion known as *Zeno's racecourse paradox*.

Zeno's Assertion:

A runner can never reach the end of a racecourse in a finite time.

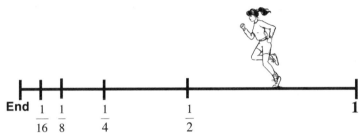

Statement	Reason
1. The runner must first pass the point $\frac{1}{2}$ located halfway between herself and the finish line before she can finish the race.	$\frac{1}{2}$ is between the runner and the finish line.
2. It will take a finite time to reach the point $\frac{1}{2}$.	It is a finite distance from 1 to $\frac{1}{2}$.
3. Once reached, there is another halfway point $\frac{1}{4}$ which the runner must reach before she can finish.	The remaining interval is divided in half.
4. There are an infinite number of such halfway points which the runner must reach and each will take a finite time.	Statements 1, 2, and 3 repeated an infinite number of times.
5. The total time for the race is infinite.	The sum of an infinite number of finite times is infinite.

But we know from experience that the runner can reach the finish line. Where is the flaw in Zeno's argument?

177

 *Infinity*

The Flaw in Zeno's Paradox

The following story may help you discover where Zeno's argument breaks down.

A famous pirate, Long John Glitter, found a cylindrical bar of gold one meter in length. To divide the bar into more manageable units he cut it in half (perpendicular to its axis of symmetry). One piece he placed in his treasure chest. The remaining piece was cut in half and one of those pieces was then placed in the chest. Again the remaining piece was cut in half and one of the pieces was placed in the chest. This process was repeated n times.

WRITE THE LENGTH OF THE nth PIECE PLACED IN THE CHEST AND THE LENGTH OF THE REMAINING PIECE.

WRITE AS A SUM OF FRACTIONS THE TOTAL LENGTH IN METERS OF THE FIRST n PIECES PLACED IN THE CHEST.

WRITE AN EXPRESSION FOR THE TOTAL LENGTH OF THE FIRST n PIECES IN THE CHEST.

SUPPOSE THIS PROCESS COULD BE CONTINUED AN INFINITE NUMBER OF TIMES. WOULD THE TOTAL LENGTH OF THE PIECES OF THE GOLD BAR EVER EXCEED 1 METER?

This example shows that the sum of an infinite number of finite quantities may indeed be finite. This calls into question the reason supporting the fifth statement in Zeno's argument above. Paradoxes such as this one prompted mathematicians to put mathematics on firmer foundations by specifying a basic set of "self-evident" truths or axioms from which all mathematical theorems could be logically and unambiguously derived. This would (hopefully) expose all hidden assumptions and remove paradoxes. Over two millenia later, deeper paradoxes arose which shook the axiomatic foundations of mathematics more profoundly than the challenges made by Zeno. One such paradox appears on page 226.

The Harmonic Series

An infinite series is a sum of a sequence with an infinite number of terms. When the sum is finite, the sequence is said to *converge*. Otherwise the series is said to diverge. For example, the series generated by the lengths of the pieces thrown into the basket (see the previous page) is given by:

$$\frac{1}{2} + \frac{1}{4} + \frac{1}{8} + \ldots + \frac{1}{2^n} + \ldots$$

It converges (to 1), but the series $1 + 1 + 1 + 1 + \ldots$ clearly does not converge.

It is important to recognize that there is a limit implicitly involved here. For example, the value of $\frac{1}{2} + \frac{1}{4} + \frac{1}{8} + \ldots + \frac{1}{2^n} + \cdots$ is defined to be the limit of the sum of the first n terms as n approaches infinity. That is, the limit of this sequence is the limit of the sequence of partial sums,

$$\frac{1}{2} \; , \; \frac{1}{2} + \frac{1}{4} \; , \; \frac{1}{2} + \frac{1}{4} + \frac{1}{8} \; , \ldots$$

The study of convergence is an important one, and often much of an undergraduate course is devoted to understanding how one can tell when a series will converge.

The harmonic series is the series: $1 + \frac{1}{2} + \frac{1}{3} + \frac{1}{4} + \ldots + \frac{1}{n} + \ldots$

For many reasons, it comes up repeatedly in many different fields of higher mathematics. There is a beautiful proof that this famous series diverges. If you have never thought about it before, you might want to ponder it yourself (or even sleep on it) before reading this remarkable argument.

The Classical Proof that the Harmonic Series Diverges

This proof is very slick. We ignore the 1, and instead sum $1/2 + 1/3 + 1/4 + \ldots$.

By the time we get to $1/2$, the partial sum is equal to $1/2$.

By the time we get to $1/4$, the partial sum is greater than $2/2$.

By the time we get to $1/8$, the partial sum is greater than $3/2$.

.

.

.

By the time we get to $1/2^n$, the partial sum is greater than $n/2$.

This means that if we are given some positive number, we can go far enough to make the partial sum bigger than this number.

The proof works by considering intervals between adjacent powers of two.

Sum between $1/2^{n-1}$ and $1/2^n$	Lower Bound	Value
$1/2$	$= 1/2$	$= 1/2$
$1/3 + 1/4$	$> 1/4 + 1/4$	$= 1/2$
$1/5 + 1/6 + 1/7 + 1/8$	$> 1/8 + 1/8 + 1/8 + 1/8$	$= 1/2$
.	.	.
.	.	.
.	.	.
$1/(2^{n-1} + 1) + \ldots + 1/(2^n - 1) + 1/2^n$	$> 1/2^n + \ldots + 1/2^n + 1/2^n$	$= 1/2$

Adding all of these together, we get:

$$\underbrace{1/2 + 1/3 + 1/4 + \ldots + 1/2^n}_{} > \overset{n \text{ terms}}{\overbrace{1/2 + 1/2 + \ldots + 1/2}} = n/2$$

That is, the partial sums increase without limit as n increases.

There are many variations on this result. Here are a few.

Variation 1

Instead of evaluating $\quad 1 + \dfrac{1}{2} + \dfrac{1}{3} + \dfrac{1}{4} + \ldots$

we can try to evaluate $\quad 1 + \dfrac{1}{2^{1.00001}} + \dfrac{1}{3^{1.00001}} + \dfrac{1}{4^{1.00001}} + \ldots$

Each term in the second series is only a tiny tiny bit smaller than the corresponding term in the harmonic series. But remarkably, the second sum converges! It converges to something very large, though. If you want to see how large and know a little about integration, you can work it out yourself, using the information that the sum is approximately the same as the following integral:

$$\int_{1}^{\infty} \frac{1}{x^{1.00001}} \, dx$$

Variation 2

Instead of evaluating $1 + \dfrac{1}{2} + \dfrac{1}{3} + \dfrac{1}{4} + \ldots$, we can try to sum the reciprocals only of those numbers with no 9's in their decimal representation. So the sum would begin:

$$1 + \frac{1}{2} + \frac{1}{3} + \ldots + \frac{1}{8} + \frac{1}{10} + \ldots + \frac{1}{18} + \frac{1}{20} + \ldots$$

It doesn't seem as though we've thrown out many numbers, but strangely, this sum also converges.

Variation 3

This time we will just sum the reciprocals of the primes, so our series will begin:

$$\frac{1}{2} + \frac{1}{3} + \frac{1}{5} + \frac{1}{7} + \frac{1}{11} + \ldots$$

Now we've thrown out a lot of numbers. There aren't that many primes — there are only 25 less than 100, 168 less than 1000, and 78,498 less than 1,000,000 — and they keep on getting rarer. But amazingly, especially in light of the previous variation, this sum *diverges*!

Rarer than primes are the so-called *twin primes*, which are any pair of prime numbers that differ by 2, such as 5 and 7 or 41 and 43. The series of the reciprocals of all twin primes converges. This tells us that twin primes are very rare indeed.

Variation 4

In *Prime Numbers in Number Theory* (p. 124), we showed using a simple method that there are an infinite number of primes. Leonhard Euler (1707-1783), the great Swiss mathematician, had another proof based on the fact that the harmonic series diverges.

The proof is subtle, but you should be able to figure it out. We use a method known as the *indirect method* — we assume the opposite of what we want to prove, and try to find a contradiction.

Imagine that there are only a finite number of primes, $P_1, P_2, P_3, \ldots, P_r$. Then every positive integer n can be expressed (uniquely) as

$$n = p_1^{m_1} p_2^{m_2} p_3^{m_3} \ldots p_r^{m_r} \text{ where the } m_i \text{ are integers.}$$

Consider the product P given by:

$$\left(1 + \frac{1}{P_1} + \frac{1}{P_1^2} + \frac{1}{P_1^3} + \ldots\right)\left(1 + \frac{1}{P_2} + \frac{1}{P_2^2} + \frac{1}{P_2^3} + \ldots\right)\cdots\left(1 + \frac{1}{P_r} + \frac{1}{P_r^2} + \frac{1}{P_r^3} + \ldots\right)$$

Since the number of primes is assumed to be finite, P is finite . (Why?)

 *Infinity*

Now imagine that you expanded out all of the brackets (in the same way that when you expand $(a + b)(c + d)$ you get $ac + bc + ad + bd$ — you get every possible product of something in the first bracket with something in the second bracket). By analogy, we would expect P to be some huge sum

$$1 + \frac{1}{p_1} + \frac{1}{p_2} + \frac{1}{p_3} + \ldots + \frac{1}{p_n} + \frac{1}{p_1 p_2} + \ldots$$

In this sum should appear every number of the form:

$$\frac{1}{p_1^{m_1} p_2^{m_2} p_3^{m_3} \ldots p_r^{m_r}}$$

We said earlier that every positive integer is of the form, $p_1^{m_1} p_2^{m_2} p_3^{m_3} \ldots p_r^{m_r}$, so P is in fact the harmonic series. Thus P is infinite.

But we said that P is finite, so we have found a contradiction. There must be an error somewhere, so our original assumption (that there are only a finite number of primes) must be false.

Hence there are an infinite number of primes.

Variation 5

We can use calculus to get more evidence that the harmonic series diverges. (As if we weren't already convinced!) Define $f(x)$ to be the following infinite series:

$$f(x) = x + \frac{x^2}{2} + \frac{x^3}{3} + \frac{x^4}{4} + \ldots$$

Notice that $f(0) = 0$, and that $f(1)$ is the harmonic series. Differentiating with respect to x, we get:
$$f'(x) = 1 + x + x^2 + x^3 + x^4 + \ldots$$

This is an infinite geometric series, so we evaluate this sum to obtain: $f'(x) = \frac{1}{1-x}$

Thus $$f(1) = f(1) - f(0) = \int_0^1 f'(x)\,dx = \int_0^1 \frac{1}{1-x}\,dx$$
$$= -\ln(1-x)\Big|_0^1$$
$$= -(-\infty)$$

(To be fair, substituting $x = 1$ into $-\ln(1-x)$ is not quite "cricket"; rather we should look at the limit of $-\ln(1-x)$ as x approaches 1 from below.)

183

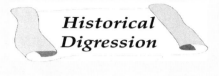

Historical Digression

The Human Computer

John von Neumann (1903-1957) was one of the preeminent mathematicians of the 20th century. At the age of thirty, he was appointed (along with Albert Einstein) as one of the first professors of the Institute for Advanced Study in Princeton. His work in mathematics and physics spanned a wide spectrum of fields and his contributions in any one of these would have won him recognition. Von Neumann's earliest contributions lay in his axiomatization of set theory; that is, he reformulated the axioms of set theory to deal with some of the problems such as Russell's Paradox, discussed in *Paradox* (p. 226). From this he proceeded to a reformulation of Quantum

John von Neumann
1903 - 1957

THE BETTMANN ARCHIVE

Mechanics. He also pioneered the development of Game Theory, the digital computer, and the study of cellular automata. Near the end of his career he worked on the atomic bomb, performing detailed calculations (often in his head) related to pressures created during implosions.

Stories about John von Neumann's remarkable capacity for mental computation abound. On one occasion he was at a cocktail party when a guest challenged him with the problem given on page 176 about the Birds and the Bee. Johnny, as he was fondly called, contemplated the problem for a second or two and then replied, "ten kilometers". The surprised interrogator retorted, "Oh, you've heard that trick before?" In typically innocent fashion, Johnny responded, "What trick? I merely summed the infinite series."

Answer to The Mathematics of the Birds and the Bee (p. 176)

One could take von Neumann's approach and calculate how far the bee travels in each direction. This requires that you find the sum of an infinite geometric series. However, there is a faster solution. The birds each take ten minutes to reach the halfway point; in this time, the bee will travel 10 km because it travels 1 km/min. That's it!

(Here is a follow-up question: How many times does the bee turn around during this journey?)

GAME THEORY REVISITED

"Come, Watson, come! The game is afoot."

— The Return of Sherlock Holmes
Sir Arthur Conan Doyle

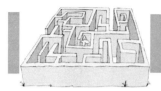

Game Theory Revisited

Sherlock Holmes' Secret Strategy: The Reykjavik Gambit

To learn how to play **Sherlock Holmes' Continental Chase**, *read* **Elementary My Dear Watson!** *(p. 74).*

If you've played around with the game awhile, you've probably convinced yourself that ENDRUN can always stay one step ahead of Holmes. But in fact, Holmes can always win. Holmes' secret trick is to start off the game by making a beeline to Reykjavik (he can get there in two moves), and then head to Edinburgh. After this, he should have no trouble chasing down ENDRUN.

Try this out and check that it works. Loosely speaking, Holmes just has to try to back ENDRUN towards a corner, usually by moving to the opposite corner of a quadrilateral from ENDRUN. For example, if Holmes were in Brussels and ENDRUN were in Venice, then Holmes should move to Munich.

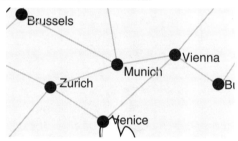

Miraculously, by going to Reykjavik Holmes has removed the potential difficulty in chasing down ENDRUN! The fifteen moves allowed Holmes are more than enough — usually, six will suffice.

This is a great game, because you can play the part of ENDRUN against an unsuspecting friend, and then (after winning repeatedly), offer to switch sides. You then appear to walk about randomly for the first few moves, and just happen to wander through Reykjavik. Around the eighth move, you suddenly seem to remember the object of the game, and quickly hunt down your bewildered opponent.

Here's a remarkable fact: Reykjavik is essential to winning the game. If you erase Reykjavik from the map (so neither Holmes nor ENDRUN may go here), then Holmes can never catch ENDRUN — even if there is no time limit and ENDRUN tries to help! (Remember — Holmes has to move to the position of ENDRUN; ENDRUN is not permitted to move to Holmes' position and turn himself in.) If you can figure out why Reykjavik is essential, then you'll have an excellent understanding of how this game works. Perhaps more important, you'll have a vital insight into other important themes addressed in this book, including parity, coloring, and graph theory. (You can take that last sentence as a hint!) When you think you've figured it out, turn to *Why Holmes' Strategy Works* (p. 189).

Game Theory Revisited

A Three-Way Duel

Why Multi-Player Games are Much Harder to Analyze

This section has three different elements mixed together: game theory, a good probability problem, and a paradox.

The Set-Up

Three players, A, B, and C, are involved in a gunfight. They take turns shooting one bullet at a time until only one player remains. C is a perfect shooter; he kills his target every time. B has an accuracy of two kills in three attempts. Finally, A only hits his target one out of every three times. To be fair, it is decided that A should begin the gunfight, to be followed by B and then C (if they are alive), and then back to A, and so on.

(Warning: Do not try this at home! These characters are trained mathematicians and use only theoretical bullets!)

The Strategy

When it is C's turn to shoot (assuming he is still alive), it is clearly in his best interest to shoot his strongest opponent. So if B is still alive, C will shoot him.

B's strategy is also clear: C is the target he must shoot first.

The Paradox

This pattern would seem to suggest that A should shoot at C if possible, and otherwise aim at B. But it turns out that A's best strategy is to deliberately miss his first shot. If A aims at C with his first shot, and B and C follow their best strategies, A will win (approximately) 31% of the time, B will win 54% of the time, and C will win 15% of the time. But if A deliberately misses on his first shot, he will win 40% of the time, B will win 38% of the time, and C will win 22% of the time.

So A's odds of winning become much better by following this counter-intuitive strategy — and he becomes the one most likely to win!

The Probability Problem

Try to work out all of the probabilities mentioned above. You can check your answers against the given percentages.

The Game-Theoretic Moral

Two-player games are much easier to analyze than multi-player games for a number of reasons. As in this example, a weak third party may initially choose not to take advantage of its position, preferring instead to let the stronger powers wear each other down. Weaker parties can hold the balance of power between two evenly matched strong opponents. Also, it can sometimes be dangerous to grow too strong too quickly, as the other players may team up if they feel threatened.

Of course, all of these reasons often make multi-player games more fun to play!

Napoleon III hopes to pick up the spoils of the Austro-Prussian War.

Courtesy of the Proprietors of *Punch*

BRITANNIA: "WELL! I'VE DONE MY BEST. IF THEY WILL SMASH
EACH OTHER, THEY MUST."
NAPOLEON: [Aside] "AND SOMEONE MAY PICK UP THE PIECES!"

Why Holmes' Strategy Works

Why Iceland is Essential

Imagine that continental drift causes Reykjavik to disappear from the playing board. Imagine also that each of the cities were colored black or white as in the map shown here. Whenever a player moves, he will move from a city of one color to a city of another color. (If you play chess, this pattern will remind you of the knight, who always moves from a square of one color to a square of the other color.) When the game begins, Holmes is in Stockholm, which is a black city. ENDRUN is in Paris, which is also a black city. Let's change the game a little. Instead of Stockholm, Holmes has to start in some arbitrary black city. Instead of Paris, ENDRUN has to start in some other black city. We'll see that even in this more general situation, Holmes can never win.

189

Holmes will start by jumping to a white city. (ENDRUN isn't there; he's in a black city.) ENDRUN will jump from a black city to a white city. Holmes will then jump from his white city to a black city, but he can't catch ENDRUN in this move, because ENDRUN is in a white city. ENDRUN will then jump from the white city to a black city. So Holmes can't catch ENDRUN in either of the first two moves of the game. But the two players are once again in a starting position, as they are both in black cities! So, by the same argument, Holmes can't catch ENDRUN in the next two moves. This can go on forever. So Holmes can't win, no matter how well he plays and how badly ENDRUN plays!

(The argument above is short, but tricky. You might want to think about it and play around on the map a little before moving on. You should think about how the argument fails if the players are allowed to go to Reykjavik.)

This game is related to many other issues such as *parity* discussed earlier (see p. 92). The game was played on a *graph* (where *graph* is meant in a technical sense), so graph theory was involved. We solved the problem by introducing a *coloring*, so this is also an example of a coloring problem (discussed in the chapter on *Chessboard Coloring*).

In particular, the graph we looked at had a very special property. The cities (called *vertices* in the lingo of graph theory) could be colored black or white in such a way that two connected vertices are always different colours. Such a graph is called a *bipartite graph*. These graphs appear remarkably often in a wide variety of situations.

Food for Thought

❶. If you look at the altered map (with Reykjavik removed), you'll notice that if you take a round trip starting anywhere and returning to the same place, the round trip must have an even number of stops (counting the beginning and end point only once). For example, one possible round trip is Madrid-Marseilles-Zurich-Venice-Vienna-Munich-Brussels-Paris-Madrid. Why is this true?

❷. (This one is tougher!) If you have a graph such that any round trip has an even number of stops, must the graph be bipartite?

How to Win at Nim

In order to win at the game of Nim, you must first know how to play! The rules were explained in The Game of Nim *(p. 66).*

The winning strategy for Nim is simple, but hard to figure out. Some calculations are required, but with a little practice you should be able to do them in your head, at least for small games.

If it is your turn to play and you are confronted with this configuration:

then you are guaranteed to lose. Less obviously, if you are confronted with the configuration:

and your opponent plays perfectly, then you are also guaranteed to lose. (Can you figure out why this is so?) We'll call configurations like this *losing configurations*. There is a quick method to check whether a given configuration is a losing configuration. Armed with this, we can play perfectly: if we are faced with a non-losing configuration, we'll see how to leave our opponent with a losing configuration. (If we are confronted with a losing configuration, then we are doomed, unless our opponent slips up and makes a non-optimal move.)

How to Evaluate a Configuration

• First of all, write the number appearing in each row in base 2 in a column (as though you were about to add them).

• Next, draw a horizontal line. Under each column, put a 1 if there are an odd number of 1's in the column, and a 0 if there are an even number in the column. This is called the *Nim-sum* of the numbers. (Computer aficionados may recognize this as XOR.) It isn't the same as adding, because we never have any "carrying".

191

As an example, we evaluate the configuration shown below.

| ———————— 1

||| ———————— 1 1

|||| ———— 1 0 0

——————
1 1 0

If there are only zeroes in the bottom row, then we have a losing configuration. (So we see that the configuration shown above isn't a losing configuration.)

This rule helps us decide how to move. If confronted with the configuration shown above, we want to leave our hapless opponent with a zero Nim-sum, so we'll need to change the number of 1's in the first and second column by altering only one number. By looking closely at the table, you will soon see that the only way to do this by *reducing* one of the numbers is to change 100 to 010 (i.e. 4 to 2), removing two toothpicks from the bottom row. Then our opponent is left with the configuration shown, which has Nim-sum zero (check it!), and therefore is a losing configuration.

| ——— 1

Try finishing this game by playing according to the strategy when it is your turn and randomly when it is your opponent's. Then you can tackle the practice questions in the *Food for Thought* section below. After you have mastered them, you will be ready to start playing (and beating) other people!

||| ——— 1 1

|| ——— 1 0

——
0 0

Food for Thought

Practice Questions

❶. What moves would you make in the following situations? Why?

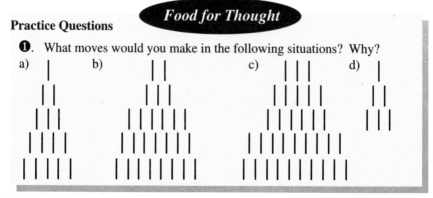

Food for Thought

❷. If there are two equal rows left in the game, who will win, the first or the second player? What course will the game take?

Further Research (You certainly don't need to solve these to play Nim perfectly!)

❸. Consider the three-pile configuration shown in the figure ($0 \le a \le 6$). For which a is this a losing position? (Make a table.) For which a is $\binom{6}{a}$ odd? Is this a co-incidence? Replace 6 with other numbers and find out! (For related discussion, see *The Triangles of Pascal, Chu Shih-Chieh, and Sierpinski*, p. 98.)

| | | | | | |

⁞ ⁞ ᵃ ⁞
⁞ ⁞ • • •⁞

⁞ ⁞ 6 - a ⁞
⁞ ⁞ • • •⁞

❹. How would one prove that the strategy described in this section will always work?

❺. If you are ready to tackle the winning strategy to another game, try these.

a) Change the rules of Nim slightly, so each player is allowed to remove as many matches as he wants from up to two piles (although at least one match must be removed). What is the winning strategy? (As in Nim, think in base 2.)

b) There are n toothpicks in a row. The first player can take up to n-1 toothpicks. The second player can take at most the number of toothpicks that the first has just taken. The first can then take at most the number of toothpicks that the second has just taken. This continues until all the toothpicks are gone. The player who takes the last toothpick wins. For which n does the first player have a winning strategy? The answer may surprise you. (As with Nim, write the number in base 2. Try to find out if the first player can force a win with $n = 1, 2, 3, 4, 5$. If the first player can force a win, what is her winning first move? Can you think of a winning first move that will work for all odd n? If so, how about numbers that are even but not divisible by 4?)

CONCEPTS IN CALCULUS

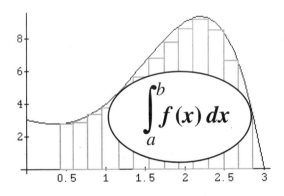

If I have seen a little further than others, it is because I have stood on the shoulders of giants.

— Sir Isaac Newton

Historical Digression

The Man Who Solved the System of the World

Johannes Kepler 1571 - 1630

Prior to the 17th century, people did not understand how the earth moved in relation to the sun. Although Copernicus had asserted in 1543 that the earth was not stationary, his theory was not generally accepted until scientists like Galileo and Tycho Brahe gathered more evidence. Furthermore, the shape of the earth's orbit and the orbits of the other planets around the sun was still unknown until Johannes Kepler enunciated his three laws of planetary motion in publications dated 1609 and 1619. Kepler's laws were empirical and it remained for scientists to explain in terms of some theory why these laws were true.

KEPLER'S THREE LAWS

1. EACH PLANET MOVES AROUND THE SUN IN AN ELLIPTICAL ORBIT. THE SUN IS LOCATED AT A FOCUS OF THE ELLIPSE.

2. EACH PLANET TRAVELS AROUND THE SUN IN SUCH A WAY THAT THE LINE JOINING THE PLANET TO THE SUN SWEEPS THROUGH EQUAL AREAS IN EQUAL TIMES.

3. IF P_1 AND P_2 ARE THE ORBITAL PERIODS OF TWO PLANETS, AND R_1 AND R_2 THE RESPECTIVE MEAN DISTANCES FROM THE SUN, THEN:

$$\left(\frac{P_1}{P_2}\right)^2 = \left(\frac{R_1}{R_2}\right)^3$$

Historical Digression

In 1663-64, Cambridge University was closed on account of the bubonic plague. Isaac Newton, who was a twenty-one year old undergraduate, was sent home to continue his studies. During that highly productive period, Newton formulated his law of universal gravitation which states that the force of attraction between any two masses is proportional to the product of their masses and inversely proportional to the square of the distance between them. At this time, Newton also developed a new branch of mathematics called *calculus*. Using the methods of calculus, he was able to deduce the three laws of Kepler from his own law of universal gravitation. He waited until 1687 to publish this work in a treatise titled *Philosophiae Naturalis Principia Mathematica*.

Isaac Newton 1642 - 1727

The scientist Pierre-Simon Laplace exclaimed,

Newton was surely the man of genius par excellence, but we must agree that he was also the luckiest: one finds only once the system of the world to be solved!

HOW DID NEWTON DEDUCE KEPLER'S SECOND LAW?

Newton reasoned as follows:

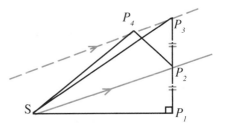

1. If there were no force, a planet would move from P_1 to P_2 in a small time, Δt. In the next time interval Δt it would move from P_2 to P_3 so that $P_1P_2 = P_2P_3$. The areas swept out in successive time intervals would be equal; that is, Area ΔSP_1P_2 = Area ΔSP_2P_3. (Why?)

2. Suppose the sun at S exerted a force on the planet in its direction. Suppose also that the "average" direction of this force over the infinitesimal time interval $2\Delta t$ were in the direction of P_2S. Then the net effect of the sun's force on the planet would be to move it parallel to P_2S away from its destination P_3 to a point P_4. The area swept out by the planet in moving from P_2 to P_4 would be Area ΔSP_2P_4. Furthermore Area ΔSP_2P_4 = Area ΔSP_2P_3. (Why?) As observed earlier, Area ΔSP_2P_3 = Area ΔSP_1P_2. Q. E. D.

197

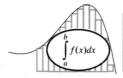

Concepts in Calculus

A Question of Continuity

Can you prove that right now, as you read this, there are two diametrically opposed points on the earth's equator that are exactly the same temperature? The answer highlights an important idea. Think about it, and then turn to page 200 for the answer.

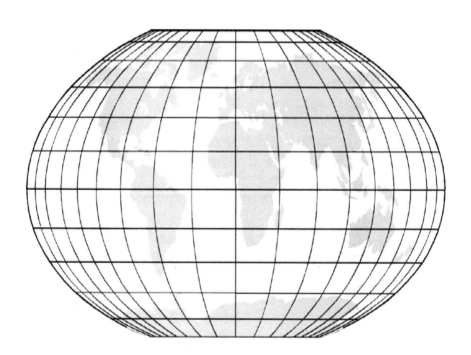

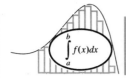

An Integration by Parts Paradox

Recall the "Integration by Parts" formula:

$$\int f(x) \frac{dg(x)}{dx} dx = f(x)g(x) - \int g(x) \frac{df(x)}{dx} dx$$

More simply: $\int f\, dg = f\, g - \int g\, df$

Let's substitute $f(x) = \dfrac{1}{x}$, $g(x) = x$ in this formula.

$$\int \frac{1}{x} dx = \left(\frac{1}{x}\right) x - \int x\, d\left(\frac{1}{x}\right)$$

$$= 1 \quad - \quad \int x \left(\frac{-1}{x^2}\right) dx$$

$$= 1 \quad + \quad \int \frac{1}{x} dx$$

Thus $\int \dfrac{1}{x} dx = 1 + \int \dfrac{1}{x} dx$, so $0 = 1$. What has gone wrong?

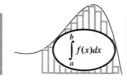

Answer to A Question of Continuity (p. 198)

It is true that, right now, there are two diametrically-opposed points on the earth's equator that are exactly the same temperature. Here's why. Imagine that a bizarre new thermometer has been invented. It doesn't display the temperature; it displays the difference between the temperature where you are and the temperature at the point on the globe diametrically opposite to you. For example, if you are standing in Quito, Ecuador and the temperature is exactly 20°C, and the temperature in the mountains of Sumatra in Indonesia diametrically opposite Quito is 25°C, then your strange thermometer will read -5.

Now, walk to some point on the equator, and look at your strange thermometer. If it reads 0, then you've found the point you're looking for. If it doesn't, then it is either positive or negative. Say it is positive. (The negative case will turn out to be exactly the same.) Then, quickly run (or swim) around the equator to the opposite side of the globe. (You'll have to do this infinitely quickly to make sure that the temperature doesn't change.) Now your strange thermometer should have the negative of the reading it had before. So if you watched it carefully while you were globe-trotting, you would have seen it change continuously from being positive to being negative. At that point, it would be zero, so that would be the spot you were looking for.

The simple idea we used was that a function (in this case, the reading on your odd thermometer) that varies continuously from one value to another must take on every value in between. This principle has the slightly pompous name of the *Intermediate Value Theorem*. Although it sounds quite trivial, it is in fact very useful. Unfortunately, in order to prove it rigorously, you will need to define exactly what you mean when you say that a function is continuous. (Waving your hands in the air doesn't count.) If you have seen one of the definitions of continuity that mathematicians use, you can try to use it to prove the *Intermediate Value Theorem*. But watch out: it is far harder than it looks!

Food for Thought

Can you prove that, right now, there are two points on the earth's equator separated by 120° that are exactly the same temperature?

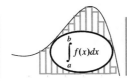

Concepts in Calculus

Dual Numbers

This section will use some elementary calculus. Complex numbers are used as an analogy, and some familiarity with them would be a big help.

Part 1: Complex Numbers

Complex numbers are numbers of the form $a + bi$, where a and b are real. They are added together in the natural way:

$$(a + bi) + (c + di) = (a + c) + (b + d)i$$

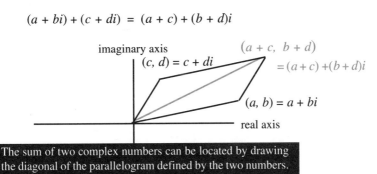

The sum of two complex numbers can be located by drawing the diagonal of the parallelogram defined by the two numbers.

They are multiplied using the fact that $i^2 = -1$:

$$(a + bi)(c + di) = a(c + di) + bi(c + di)$$
$$= ac + (ad + bc)i + bd\, i^2$$
$$= (ac - bd) + (ad + bc)i$$

Division can be worked out with a little cleverness. The top and bottom of a fraction are multiplied by the *conjugate* of the denominator, and everything magically works out:

$$\frac{a+bi}{c+di} = \frac{(a+bi)(c-di)}{(c+di)(c-di)}$$
$$= \frac{(ac+bd)+(bc-ad)i}{c^2+d^2}$$
$$= \frac{ac+bd}{c^2+d^2} + \frac{bc-ad}{c^2+d^2}i$$

(How are the line segments corresponding to two complex numbers related in magnitude and direction to the line segments corresponding to their product and quotient?)

201

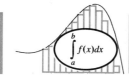

Concepts in Calculus

Dual Numbers

Part 2: From Complex to Dual Numbers

The *dual numbers* are similar to complex numbers: they are numbers of the form $a + b\varepsilon$, where a and b are real. (The symbol ε is an *epsilon*. Epsilon is the fifth letter in the Greek alphabet.) Dual numbers are added in the natural way, so $(3+4\varepsilon) + (5-2\varepsilon) = (8+2\varepsilon)$, and they are multiplied using the fact that $\varepsilon^2 = 0$. (But ε isn't zero! Yes, this is strange!)

For example,
$$(3 + 4\varepsilon)(5 - 2\varepsilon) = 3(5 - 2\varepsilon) + 4\varepsilon(5 - 2\varepsilon)$$
$$= 15 - 6\varepsilon + 20\varepsilon - 8\varepsilon^2$$
$$= 15 + 14\varepsilon$$

You should try an easy example to make sure you understand how multiplication works. For example, you can check that $(2 + \varepsilon)(6 - \varepsilon) = 12 + 4\varepsilon$, or that $(a + b\varepsilon)(c + d\varepsilon) = ac + (ad + bc)\varepsilon$.

The method for division is similar to that for complex numbers: we multiply the top and bottom of a fraction by the conjugate of the denominator. For example,
$$\frac{4+2\varepsilon}{2+7\varepsilon} = \frac{(4+2\varepsilon)(2-7\varepsilon)}{(2+7\varepsilon)(2-7\varepsilon)}$$
$$= \frac{8 - 24\varepsilon}{4}$$
$$= 2 - 6\varepsilon$$

(Work through the algebra yourself — a few of the steps have been deliberately omitted.) There's a catch, though. With the real numbers, you can never divide by zero. With the dual numbers, you can never divide by $c + d\varepsilon$ when $c = 0$. (What happens in this case?)

If complex numbers are old hat to you, you will become comfortable with dual numbers quite quickly. Once you feel comfortable with computations using dual numbers, you can move on to more interesting applications.

202

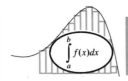

Part 3: Dual Numbers and Calculus

We will now look at more complicated versions of dual numbers. They will still be of the form $a + b\varepsilon$, but this time a and b will be functions of x. One (odd) example is:

$$(3 + 2x) + \varepsilon (\sin x).$$

Next, we will look at a special set of dual numbers, which we will call *good* dual numbers. These are dual numbers of the form

$$f(x) + f'(x)\, \varepsilon.$$

(You might prefer the notation $f(x) + \varepsilon\, \dfrac{df(x)}{dx}$; do whatever you like best.)

For example, $(2x^3) + \varepsilon(6x^2)$ is a good dual number, as is $\sin x + \varepsilon \cos x$. So is 3 (which is just $3 + 0\varepsilon$). But $(3x^2) + \varepsilon(3x)$ is *not* a good dual number.

The strange thing is that you can add, subtract, multiply, and divide good dual numbers, and the answers are still good. For example,

$$(f(x) + \varepsilon f'(x))(g(x) + \varepsilon g'(x)) = f(x)g(x) + \varepsilon(f(x)g'(x) + f'(x)g(x))$$

Check the algebra!

And, as $\dfrac{d\big(f(x)g(x)\big)}{dx} = f'(x)g(x) + f(x)g'(x)$, this is still a *good dual number.*

In fact, if you have trouble remembering the Product Rule, this is a good way of figuring it out on the spot. In much the same way that *de Moivre's Theorem*[1] can help you remember trigonometric identities, dual numbers can help you remember identities in calculus.

Let's now take the square of $f(x) + \varepsilon f'(x)$. We get

$$(f(x) + \varepsilon f'(x))^2 = (f(x))^2 + \varepsilon\,(2f(x)f'(x)).$$

Sure enough, $\dfrac{d\big(f(x)^2\big)}{dx} = 2f(x)f'(x).$

[1] You can read about this theorem in *i^i and Other Improbabilities* (p. 208).

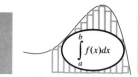
The time has come for you to explore a bit yourself. (If you don't find it confusing, you might want to replace $f(x)$ with f. For example, our squaring calculation becomes a little shorter: $(f + \varepsilon f')^2 = f^2 + \varepsilon (2ff')$.) Start simple; try cubing, and then taking fourth powers. Try taking n^{th} powers.

Then, try to re-derive the Quotient Rule: work out $(f + \varepsilon f')/(g + \varepsilon g')$ using the method of dividing dual numbers described earlier and see what you get.

Food for Thought

If you're willing to get your hands dirty, you can dig into these hornet's nests. But keep in mind that, as with actual mathematical research, there is a lot to do here. There is no point at which you can say "I'm done, and I've completely solved the problem." Each small victory opens up new questions.

❶. You can try to prove the Chain Rule. To get started, you can think about the following questions. If f is a function, instead of plugging in x, what happens if you plug in $(x + \varepsilon)$? Your answer should be another dual number. (As always, start playing with easier cases: see what happen when f is a polynomial, or even just x^2 or x^3.) What happens if you plug in $(x + 3\varepsilon)$? How about $(x^2 + 2x \varepsilon)$?

❷. *Hyper-dual numbers* are defined as follows. Instead of i's (which satisfy $i^2 = -1$) or ε's (which satisfy $\varepsilon^2 = 0$), we have e's which satisfy $e^3 = 0$. A *hyper-dual* number is of the form $a + b\,e + c\,e^2$. Figure out how to multiply them. Can you figure out how to divide them? Now, try to think of a definition of a *good hyper-dual number*. $(f + e f' + e^2 f'')$ won't

do — why not? Try $\left(f + e f' + \dfrac{e^2}{2} f'' \right)$. Why is this better?

❸. (Continuing #2, getting more and more abstract and tricky!) Could you define hyper-hyper-dual numbers, using $z^4 = 0$? What would you conjecture would be good hyper-hyperdual numbers? Can you guess how this might generalize?

If you manage to get some partial ideas on any of these questions, congratulations! You are doing honest-to-goodness mathematical research!

$\mathscr{P}$ersonal $\mathscr{P}$rofile

$\mathscr{A}$ $\mathscr{B}$ig '$\mathscr{Y}$es!!!' $\mathscr{G}$oes $\mathscr{T}$hrough my $\mathscr{H}$ead

Vin de Silva
(United Kingdom)
Born Dec. 2, 1971

Born in Colombo, Sri Lanka, Vin de Silva moved to London with his parents in 1973. As a child, his interests ranged from rubbish trucks to dinosaurs, with mathematics lying somewhere in between.

In fact, Vin's eventual career in mathematics began rather inauspiciously. When he was about three, he was shouted at by a salesclerk in the mathematical section of *Foyles*, the largest bookstore in the country, for dropping a math book on the floor. Despite his ordeal, Vin began to show an aptitude for arithmetic in primary school — something which his father encouraged by helping him with his sums at home.

At Dulwich College, a boys' public school, he continued to do well at math, but also hugely enjoyed classes in French, Latin, Greek, and other languages. His interests included chess, bridge, origami, "dino-mania", and writing.

Over the years, Vin gradually developed a love for music through the study of the violin. He took eight years of lessons, but initially saw them only "as an exercise in mind-to-finger dexterity", and so he played very badly. When his first teacher left after five years, he continued to play, but only out of habit. Fortunately, his second teacher proved to be a great inspiration; by the time Vin was in his final year at Dulwich, he was Leader of the Orchestra, an honor which had been unthinkable a few years earlier. He learned a very valuable lesson from his teacher: "If you work at things carefully then they will improve and perhaps become very much better than you could have imagined at the start."

The lesson came just in time to help Vin realize that he could work at his mathematics and go a long way. Until then, math had been "instant gratification or nothing" — he often felt impatient with problems and pondered them for only a few moments

before he looked at the answers. In Third Form, five years before the end of secondary school, a far-sighted teacher entered Vin in the national mathematical contest. Although geared largely toward students in their final two years, Vin came fifth in his school. This success made Vin sit up and decide to be a mathematician. With support from Paul Woodruff, his form master for his last two years at Dulwich, he continued to improve dramatically. In his final two years of secondary school, he did very well on the *British Mathematical Olympiad* (BMO). His ranking qualified him to be a member of the British team to the *International Mathematical Olympiad* (IMO) in Braunschweig, Germany in 1989. There he won a Bronze Medal. He followed this performance with a Gold Medal win in Beijing, China in 1990.

After Dulwich, Vin studied mathematics at Trinity College, Cambridge, while simultaneously earning a diploma in Russian. Since then, he has been pursuing a D.Phil at Oxford University under Simon Donaldson, a winner of the prestigious Fields Medal (the mathematician's equivalent of a Nobel Prize).

Vin has also continued to work with talented high school students. He sits on the BMO Committee, and has helped out at the *IMO Training Selection Weekends* at Trinity College, Cambridge. As a result, in 1994 he was invited to be the Deputy Leader of the British Team, and traveled to Hong Kong for the IMO.

For Vin, the attraction of mathematics is clear: "The sensation of suddenly seeing right through something and understanding it completely is incredible. A big 'Yes!!!' goes through my head."

COMPLEX NUMBERS

$$e^{i\vartheta} = \cos\vartheta + i\sin\vartheta$$

$$\sqrt{-1}$$

$$e^{i\pi} = -1$$

*The moving power of mathematical invention
is not reasoning but imagination.*

—A. De Morgan

$\sqrt{-1}$ *Complex Numbers*

i^i *and Other Improbabilities*

For this section, some familiarity with complex numbers will be essential. Prior knowledge of de Moivre's Theorem would also be helpful. This section is quite difficult, but a lot of amazing things happen by the end.

You can use your calculator to compute the value of a^b, where a and b are real numbers and a is positive. For example, you can easily check that $2^7 = 128$ and $3^{0.5} \approx 1.732050808$. But as soon as you leave the positive reals, troubles arise. If you ask your calculator to compute the square root of -1 (in other words, $(-1)^{0.5}$), unless it is very sophisticated it will give you an error. But (if you've seen complex numbers) you know that both i and $-i$ are "square roots" of -1, in that if you square either of them you'll get -1.

With the situation this complicated already, it may seem outrageous to ask about the value of i^i. It may seem more outrageous still to claim that, in some reasonable sense, $i^i = e^{-\pi/2} \approx 0.200787958$. And it must seem over the top to finally claim that i^i is *also* (approximately) equal to 111.31777, 59609.741, and even more numbers. But with a little ingenuity, creativity, and a result known as *de Moivre's Theorem* you can discover this, and much more.

de Moivre's Theorem

$$e^{i\theta} = \cos\theta + i\sin\theta$$

Here theta must be in radians, not degrees.[1]

For example, if we let $\theta = \pi$, we get

$$e^{i\pi} = \cos\pi + i\sin\pi$$
$$= \cos 180° + i\sin 180°$$
$$= -1$$

Skip this step if you're comfortable thinking in radians

so,

$$e^{i\pi} + 1 = 0.$$

How can it be that a transcendental number when raised to a power which is not only transcendental but imaginary, can be added to 1 to yield absolutely nothing?!!

---

[1] Radians and degrees are different units of angular measure, related by the formula: π radians = 180°.

The Oddly Polyglot Statement $e^{i\pi} + 1 = 0$

The famous equation $e^{i\pi} + 1 = 0$ makes an interesting appearance in James Gleick's *Genius*, a popular biography of the great physicist Richard Feynman. In 1944, at the age of 25, Feynman delivered a lecture to the world's most distinguished physicists who were gathered together in Los Alamos to develop an atomic bomb. In a letter to his mother, Feynman reported on this lecture, titled *Some Interesting Properties of Numbers*, which he had delivered as a prelude to a new formulation of Quantum Physics. Gleick describes Feynman's report to his mother:

Richard Feynman
1918-1988

Courtesy of the Archives, California Institute of Technology

...Now he posed one more question, as fundamental as all the others, yet encompassing them all in the round recursive net he had been spinning for a mere hour: To what power must e be raised to reach i? (They already knew the answer, that e and i and π were conjoined as if by an invisible membrane, but as he told his mother, "I went pretty fast and didn't give them a hell of a lot of time to work out the reason for one fact before I was showing them another still more amazing.") He now repeated the assertion he had written elatedly in his notebook at the age of fourteen, that the oddly polyglot statement $e^{i\pi} + 1 = 0$ was the most remarkable formula in mathematics. Algebra and geometry, their distinct languages notwithstanding, were one and the same, a bit of child's arithmetic abstracted and generalized by a few minutes of the purest logic. "Well," he wrote, "all the mighty minds were mightily impressed by my little feats of arithmetic." [1]

---

[1] James Gleick, *Genius*. (New York: Vintage Books, 1992), p. 183.

$\sqrt{-1}$ *Complex Numbers*

If you've seen the power series expansions of e^x, cos x, and sin x, then de Moivre's Theorem shouldn't surprise you.

$$e^x = 1 + x + \frac{x^2}{2!} + \quad \ldots \quad + \frac{x^j}{j!} + \quad \ldots \quad = \sum_{j=0}^{\infty} \frac{x^j}{j!}$$

$$\cos x = 1 - \frac{x^2}{2!} + \frac{x^4}{4!} - \quad \ldots \quad + (-1)^j \frac{x^{2j}}{j!} + \quad \ldots \quad = \sum_{j=0}^{\infty} (-1)^j \frac{x^{2j}}{(2j)!}$$

$$\sin x = x - \frac{x^3}{3!} + \frac{x^5}{5!} - \quad \ldots \quad + (-1)^j \frac{x^{2j+1}}{(2j+1)!} + \quad \ldots \quad = \sum_{j=0}^{\infty} (-1)^j \frac{x^{2j+1}}{(2j+1)!}$$

These power series require x to be in radians. (Incidentally, this is one reason why radians are in some sense more natural units for sin, cos, and other trigonometric functions. The power series would get a lot uglier if the units were in degrees. This explains why mathematicians find e^x more natural than 10^x. So once again e and π seem closely related.) With a little algebraic manipulation you will see that e^{ix} and cos $x + i$ sin x have the same power series expansions, "proving" de Moivre's Theorem. The real power of de Moivre's Theorem becomes evident when we use it to derive a plethora of trigonometric identities, some of which are otherwise difficult to derive.

This theorem enables us to convert any complex number given in Cartesian form, such as $z = R\cos \theta + iR\sin\theta$, into *polar form*, $z = Re^{i\theta}$. In this form, R is the magnitude of the complex number and θ is its argument. Multiplication and division of complex numbers in polar form is simply a matter of applying

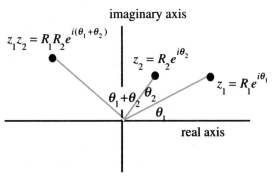

the exponent laws. Hence we can easily verify that if z_1 and z_2 are two complex numbers, then $|z_1 z_2| = |z_1||z_2|$ and $\arg(z_1 z_2) = \arg(z_1) + \arg(z_2)$. The corresponding identities hold for the quotient of two complex numbers.

$\sqrt{-1}$ *Complex Numbers*

Addition Formulas for Sine and Cosine

In what follows we use de Moivre's Theorem to derive the addition and subtraction formulas for cosine and sine. You might find this a convenient trick to remember these formulas. This technique can be applied to many common trigonometric identities.

From de Moivre's Theorem, $\qquad e^{i(x+y)} = \cos(x+y) + i\,\sin(x+y)$

But $\qquad e^{i(x+y)} = e^{ix}\,e^{iy}$ (by standard rules of exponents)

$\qquad\qquad\qquad = (\cos x + i\,\sin x)\,(\cos y + i\,\sin y)$

$\qquad\qquad\qquad = (\cos x \cos y - \sin x \sin y) + i\,(\cos x \sin y + i \cos y \sin x)$

Comparing real parts, we get

$$\boxed{\cos(x+y) = \cos x \cos y - \sin x \sin y}$$

and comparing imaginary parts, we find

$$\boxed{\sin(x+y) = \cos x \sin y + \cos y \sin x.}$$

Two complex expressions are equal if and only if the real parts are equal and the imaginary parts are equal. That is, $f_1(x) + i\,g_1(x) = f_2(x) + i\,g_2(x)$ if and only if $f_1(x) = f_2(x)$ and $g_1(x) = g_2(x)$.

Food for Thought

❶. Find $\cos 4\theta$ and $\sin 4\theta$ in terms of $\cos \theta$ and $\sin \theta$.

❷. $e^{i\theta}\,e^{-i\theta} = e^0 = 1$. Expanding the left-hand side and using $\cos(-\theta) = \cos \theta$ and $\sin(-\theta) = -\sin\theta$, rediscover a well-known identity.

In an earlier section, we proved that arctan $1/2$ + arctan $1/3$ = arctan 1. You can easily prove it another way using the idea of an *argument* of a complex number.
❶. First verify the equation $(2 + i)(3 + i) = 5 + 5i$.
❷. Then take arguments of both sides. What identity does $(1 + 2i)(1 + 3i) = -5 + 5i$ correspond to?

In 1706, John Machin (1680-1752), Professor of Astronomy in London, computed the first 100 digits of π using the identity

$$\pi / 4 = 4 \text{ arctan } 1/5 - \text{arctan } 1/239 \quad \text{(in radians)}$$

Can you prove this using complex numbers? It was also in 1706 that the symbol π was first used to denote the ratio of the circumference of a circle to its diameter.[1]

---

[1]Petr Beckmann, *A History of Pi*. (Boulder, Colorado: The Golem Press, 1971).

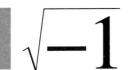 *Complex Numbers*

Computing i^i

Now we're ready to make our initial computation of i^i.

$$e^{\pi i/2} = \cos\frac{\pi}{2} + i\sin\frac{\pi}{2}$$
$$= \cos 90° + i\sin 90°$$
$$= 0 + i$$
$$= i$$

Therefore,
$$i^i = \left(e^{\pi i/2}\right)^i = e^{(\pi i/2)i} = e^{-\pi/2}$$
$$\approx 0.20787958$$

Fantastic! For some strange reason, i^i is a real number!
But...

$$e^{-3\pi i/2} = \cos\frac{-3\pi}{2} + i\sin\frac{-3\pi}{2}$$
$$= \cos(-270°) + i\sin(-270°)$$
$$= \cos 90° + i\sin 90°$$
$$= i$$

so

$$i^i = \left(e^{-3\pi i/2}\right)^i = e^{(-3\pi i/2)i} = e^{3\pi/2}$$
$$\approx 111.31777$$

Oh no! We've found that i^i is actually *another* number!

(Can you show, by a similar argument, that $i^i = e^{(2k\pi - \pi/2)}$ for any integer k?)

To see what has gone wrong, as well as what has gone right, we will have to delve a little deeper.

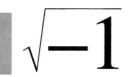

Roots of Unity

The crux of our problem is that we can express i in many different ways:

$e^{\frac{-3\pi i}{2}}, e^{\frac{\pi i}{2}}, e^{\frac{5\pi i}{2}}, e^{\frac{9\pi i}{2}}, \ldots$ This is true of every complex number. For example, 1 can be expressed as e^0, but it can also be expressed as $e^{2\pi i}$, $e^{4\pi i}$, $e^{6\pi i}$, ..., and $e^{-2\pi i}$, $e^{-4\pi i}$, ... (Don't take my word for it — use de Moivre's Theorem to check it out!)

Let's take the square root of 1. ("But surely the answer is 1!" you cry. Not so fast!)

We know that $1 = e^{2\pi k i}$ where $k = ..., -2, -1, 0, 1, 2, ...$

So $\qquad 1^{\frac{1}{2}} = \left(e^{2\pi k i}\right)^{\frac{1}{2}}$

$$= e^{\pi k i}$$

$$= \cos 180k° + i \sin 180k°$$

$$= \begin{cases} 1 \text{ if } k \text{ is even} \\ -1 \text{ if } k \text{ is odd} \end{cases}$$

Unlike the "usual" square root, which gives us one answer, this method gives us both roots of 1!

Let's try that again, but this time let's find the cube root of 1:

$$1^{\frac{1}{3}} = \left(e^{2\pi k i}\right)^{\frac{1}{3}} = e^{\frac{2}{3}\pi k i}$$

$$= \begin{cases} 1 & \text{when } k \text{ is divisible by 3} \\ \cos 120k° + i \sin 120k° & \text{when } k \equiv 1 (\text{mod } 3) \\ \cos 240k° + i \sin 240k° & \text{when } k \equiv 2 (\text{mod } 3) \end{cases}$$

$$= \begin{cases} 1 & \text{when } k \text{ is divisible by 3} \\ \dfrac{-1}{2} + \dfrac{\sqrt{3}}{2} i & \text{when } k \equiv 1 (\text{mod } 3) \\ \dfrac{-1}{2} - \dfrac{\sqrt{3}}{2} i & \text{when } k \equiv 2 (\text{mod } 3) \end{cases}$$

$k \equiv 1 (\text{mod } 3)$ just means that $k - 1$ is divisible by 3 (or that k leaves a remainder of 1 upon division by 3). Similarly, $k \equiv 2 (\text{mod } 3)$ means that $k - 2$ is divisible by 3.[1]

---

[1] Modular arithmetic was used earlier in *Divisibility Rules* (p. 22).

This should give us the three cube roots of 1. Try and work out $\left(\dfrac{-1}{2} + \dfrac{\sqrt{3}}{2}i\right)^3$ and see what you get.

Food for Thought

Can you find the three cube roots of -1? How about the four fourth roots of 2? How about the ith *power* of 1? (Watch out: there are a lot! And don't expect them to be real!)

Back to a^b and i^i

If b is an integer, then a^b has a single value. To find it, we just need to multiply a together b times.

If b is rational but not an integer (1/2 or 1/3 for example), things get more complicated. a^b may take on any of a finite number of values. (If $b = 2/5$, how many values will a^b take on?)

But if b isn't even rational (for example, $\sqrt{2}$, π, e, or i) then a^b can actually take on an infinite number of values. So all of our computations of i^i were in fact correct.

Strange but true!

Conclusion

There is a lot going on here, so don't be surprised if it takes you a long time to absorb everything that happened. There's even more going on beneath the surface. (For example, whenever you think of an exponential equation of the form $e^x = y$, there's always a "*ln*" lurking nearby: $x = \ln y$. Just as something unusual is going on with the exponential for complex numbers, something odd happens with ln.)

$\sqrt{-1}$ *Complex Numbers*

Three Impossible Problems of Antiquity

As we saw earlier in *The Pythagorean Pentagram, The Golden Mean, and Strange Trigonometry* (p. 154), the ancient Greeks saw mathematics as intrinsically related to beauty and perfection. According to Plato, the only perfect geometrical figures are the straight line and the circle. For this reason, the Greeks used only two instruments for performing geometrical constructions, the straightedge and the compasses. (A straightedge is a ruler without any markings indicating length.) With these two instruments, they were able to perform many different constructions. For example, parallel lines can be drawn, angles bisected, and segments divided into any number of equal segments. Given any polygon, it is possible (using clever methods) to construct a square of equal area, or of twice the area. The Greeks could do all this and much more. But there were three tasks that they were unable to do using these tools: *doubling the cube*, *squaring the circle*, and *trisecting an angle*.

It is said that after a plague had wiped out much of the population of Athens, the city sent a delegation to the oracle of Apollo at Delos to ask how the plague could be stopped. The oracle gave a characteristically cryptic reply: the cubical altar to Apollo must be doubled. The Athenians dutifully doubled the dimensions of the altar, but to no avail; the volume had increased in size by a factor of eight, not two. This is the legendary source of the "Delian problem" (the doubling of the cube), that is, given a line segment, to construct (with straightedge and compasses only) a segment $\sqrt[3]{2}$ times greater.

The problem of squaring the circle has a similar flavor. Given a line segment, a circle can be constructed with the segment as radius. Is it possible (using only a straightedge and compasses) to construct a square with the same area as this circle? Equivalently, given a line segment, is it possible to construct a segment $\sqrt{\pi}$ times greater?

The third problem is slightly different. Given an arbitrary angle, there is a well-known of method of bisecting it (dividing it in two) using just straightedge and compasses. Is there a method to *trisect* a general angle?

The ancient Greeks expended an inordinate amount of energy and ingenuity trying to solve these three problems, but to no avail. We now know why — all three constructions are impossible. It isn't that no one has been clever enough to find a

$\sqrt{-1}$ *Complex Numbers*

construction — it has been mathematically proven that no such construction exists. Even today, many people send "solutions" of these classical problems to university math departments. Thanks to the mathematical proof, those in the know realize (even without looking at them) that the solutions *can't* be right.

How do mathematicians know this? Strangely enough, the proof of this geometric fact uses an algebraic argument. You can denote each point in the plane with an ordered pair of real numbers (x, y). These are called *Cartesian co-ordinates*, after René Descartes (1596-1650), a French mathematician and philosopher who is known to many because of his famous statement, "Cogito ergo sum" —"I think therefore I am".

The co-ordinates (x, y) can also be represented by the complex number $x + yi$, so that, for example, $2 + 3i$ represents the point $(2, 3)$ and $-i$ represents the point $(0, -1)$. If you are given the location of 0 in the plane (that's $0 + 0i$ and 1 (that's $1 + 0i$) and a ruler and compasses, then the points you can construct using these tools and these two given points are a subset of the complex numbers known as the *constructible numbers*. It turns out that the constructible numbers are those numbers that can be expressed using complex numbers of the form $m + in$ (where m and n are integers) and the operations of addition, subtraction, multiplication, division, and square roots. For example, you can construct the point at $87/5$ and the point at i as well as the point at

$$\frac{-17\sqrt{6 + \sqrt{4 + 2i}}}{\sqrt{-3 + \dfrac{4}{3}}} \; .$$

The proof uses what is called "Galois theory", named after Évariste Galois. Now, doubling the cube means constructing the number $\sqrt[3]{2}$. Using Galois theory, it is possible to prove that you cannot express $\sqrt[3]{2}$ in terms of numbers of the form, $m + in$ (where m and n are integers), and addition, subtraction, multiplication, division, and square roots, so doubling the cube is impossible.

Squaring the circle means constructing the number $\sqrt{\pi}$. It turns out that all constructible numbers are zeroes of polynomials with integer co-efficients (they are *algebraic*) while $\sqrt{\pi}$ isn't (it is *transcendental*; this follows from the fact that π is transcendental), so $\sqrt{\pi}$ isn't constructible. (Algebraic and transcendental numbers are described further in *Numbers, Numbers, and More Numbers* on page 118 and *The Existence of Transcendental Numbers* on page 237.)

$\sqrt{-1}$ *Complex Numbers*

Showing the impossibility of the trisection of the angle is trickier still. We will show that you cannot trisect the 60° angle. For if you could, you could construct the number cos 20° starting with just $0 + 0i$ and $1 + 0i$ as follows. First construct a 60° angle, as shown in the diagram. Then trisect the angle to get a 20° angle. Intersect the resulting ray with a circle of radius 1. This gives the point (cos 20°, sin 20°). Finally, project to the *x*-axis to get a segment of length cos 20°.

It turns out that cos 20° (which is a solution of the cubic equation $8c^3 - 6c - 1 = 0$) cannot be expressed in terms of sums, differences, products, or quotients of numbers of the form,

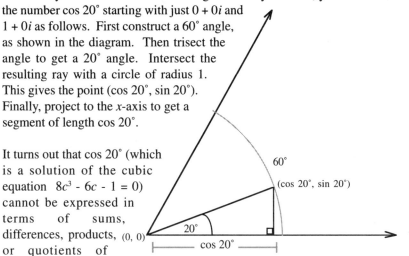

$m + in$ or $\sqrt{m+ni}$ where *m* and *n* are integers, so you can't construct it. But if you can't even trisect the 60° angle, there obviously can't be a method of trisecting a general angle!

These three results are quite subtle, and as you can see, a few paragraphs cannot hope to do them justice. But I hope you now have a feel for how they work, and how complex numbers have unusual applications in geometry.

$\sqrt{-1}$ *Complex Numbers*

Galois Theory

Galois Theory is among the most beautiful mathematics taught to undergraduates. Some of the ideas of Galois theory come up in this problem.

Problem. Prove that $11 + 14\sqrt{2}$ cannot be expressed as the square of $(a + b\sqrt{2})$, where a and b are rational numbers.

Solution. If $11 + 14\sqrt{2} = (a + b\sqrt{2})^2 = (a^2 + 2b^2) + (2ab)\sqrt{2}$, then $11 = a^2 + 2b^2$ and $14 = 2ab$, so
$$11 - 14\sqrt{2} = (a^2 + 2b^2) - (2ab)\sqrt{2} = (a - b\sqrt{2})^2$$

But the right side is the perfect square of a real number, which must be at least zero, and the left side is negative, so this is impossible.

> In order to make this proof complete, we need to know that if $11 + 14\sqrt{2} = c + d\sqrt{2}$ and c and d are rational, then $c = 11$ and $d = 14$. Can you prove this using the fact that $\sqrt{2}$ is irrational? (The irrationality of $\sqrt{2}$ was shown in *First Steps in Number Theory* on page 120.)
>
> *Hint*: rephrase the equation as: $(11 - c) = (d - 14)\sqrt{2}$.

218

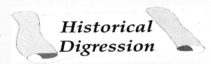

Historical Digression

Pistols at Dawn — The Teenager who Launched Abstract Algebra

Évariste Galois
1811 - 1832

Évariste Galois was born near Paris on Oct. 25, 1811 during the turbulent Napoleonic years. His family was strongly political, and his father, a Republican, was village mayor. Being a precocious child, Galois was bored in school, and despite showing great promise in Latin, he was held back for a year. To alleviate the tedium of his classes, he started reading mathematics, and by the age of fifteen he was reading material intended for professional mathematicians.

Galois hoped to enrol at the École Polytechnique, the breeding ground for French mathematics. Ignoring his teachers' advice to work systematically, he took the competitive entrance exam without adequate preparation and failed. A few days after his father's suicide in 1829, he took the examination a second time, and failed again.

Undaunted, Galois continued his education at another school, where he showed tremendous promise. But misfortune continued to haunt him. At seventeen, the young Galois presented a paper to the prestigious Academy of Sciences, only to have it lost by an Academy official, Augustin-Louis Cauchy. Then, in February 1830, Galois submitted his work to be judged for the Grand Prize in Mathematics, the ultimate mathematical honor. J.-B. Fourier, the secretary of the Academy, took the manuscript home, but died before reading it. Galois' brilliant research could not be found among Fourier's papers.

Thoroughly disillusioned by his misfortunes, Galois took part in the 1830 Revolution. In 1831, he was arrested, and spent time in jail. Soon after, he began a love affair with a mysterious Mademoiselle Stéphanie D. Subsequently he was challenged to a duel ostensibly because of the young woman, but perhaps because of his politics. He spent the eve of the duel, feverishly scribbling his mathematical discoveries. The next morning, May 30, 1832, he met his adversary in a duel with pistols. Galois was shot in the stomach and died a day later, aged twenty!

Only after his tragic death was the enormous importance of his work realized. His research set the stage for the results that have been discussed in this section, but his ideas are far more wide-ranging. Even today, *Galois theory* is an integral part of much of mathematics. One can only wonder what Galois might have done had he been given a few more years.

219

$\sqrt{-1}$ *Complex Numbers*

Using Complex Numbers in Geometry

Although complex numbers are extremely useful in many fields of mathematics, you wouldn't think that they could possibly be of any use in geometry. Surprisingly, they can be used to prove unexpected results such as the following:

Theorem

> Consider a regular n-gon inscribed in a unit circle with vertices $P_0, P_1, ..., P_{n-1}$. Then
>
> $$\left|P_1 P_0\right|\left|P_2 P_0\right|\left|P_3 P_0\right| \quad ... \quad \left|P_{n-1}P_0\right| \; = \; n.$$

Proof.

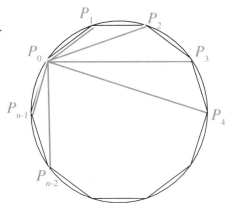

In *Three Impossible Problems of Antiquity* (p. 215), we saw how complex numbers can be represented by points in the plane. For example, $3 + 4i$ is represented by the point $(3, 4)$. Consider the n points in the complex plane (also known as the *Argand plane*) corresponding to the nth roots of 1: w^0, $w, w^2, ..., w^{n-1}$ (where w is a primitive n^{th} root of 1). These points are the vertices of a regular n-gon inscribed in a unit circle.

We can let the points $P_0, P_1 ..., P_{n-1}$ be at these points, where P_0 is at the 1 (which we'll write as "$P_0=1$"), and in general P_j is at w^j. Then

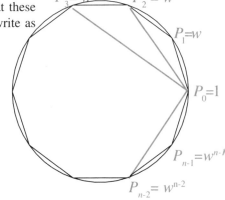

$$\left|P_1 P_0\right|\left|P_2 P_0\right|\left|P_3 P_0\right| \quad ... \quad \left|P_{n-1}P_0\right|$$

$$= \; \left|1-w^1\right|\left|1-w^2\right|\left|1-w^3\right|...\left|1-w^{n-1}\right|$$

$$= \; \left|(1-w)(1-w^2)(1-w^3)...(1-w^{n-1})\right|$$

$$= \; \left|f(1)\right|$$

where $f(x)$ is defined by:

$$f(x) = \frac{(x-1)(x-w)(x-w^2)(x-w^3)...(x-w^{n-1})}{(x-1)}$$

But $x^n - 1 = (x - 1)(x - w)(x - w^2)...(x - w^{n-1})$ (as both sides of the equation have the same roots and leading coefficient 1)

so $f(x) = \dfrac{x^n - 1}{x - 1}$

$$= 1 + x + x^2 + ... + x^{n-1}$$

Substituting $x = 1$, we get: $f(1) = 1 + 1 + 1 + ... + 1 = n$

Therefore, $\left|P_1 P_0\right|\left|P_2 P_0\right|\left|P_3 P_0\right| \quad ... \quad \left|P_{n-1} P_0\right| = |n| = n$ Q. E. D.

Now that you've seen how complex numbers work in geometry, you can try to prove the following theorem yourself:

Theorem.

Consider a regular n-gon inscribed in a unit circle with vertices P_0, P_1,..., P_{n-1} and centre O, and a point Q somewhere on the ray $O P_0$. Then

$$\left|P_0 Q\right|\left|P_1 Q\right| \quad ... \quad \left|P_{n-1} Q\right| = \left|\,|OQ|^n - 1\right|.$$

$\sqrt{-1}$ Complex Numbers

Food for Thought

Many other results can also be proved using complex numbers, although their proofs are too long to get into here. (Warning: If you want to tackle them yourself, prepare for a challenge!)

❶. Choose some random angle θ (degrees). Using your calculator, evaluate

$$\cos\theta + \cos(\theta + 120°) + \cos(\theta + 240°)$$
and
$$\sin\theta + \sin(\theta + 120°) + \sin(\theta + 240°)$$

What do you notice? Why is this true? Can you generalize it?

Hint: Evaluate
$$\cos\theta + \cos(\theta+72°) + \cos(\theta+144°) + \cos(\theta+216°)+\cos(\theta+288°).$$

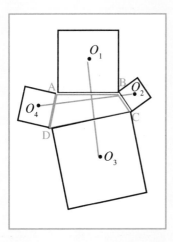

❷. Squares (with centers O_1, O_2, O_3, O_4, respectively) are constructed externally on the sides of a given convex quadrilateral ABCD. Then $|O_1O_3| = |O_2O_4|$ and O_1O_3 is perpendicular to O_2O_4.

❸. A hexagon ABCDEF is inscribed in a unit circle, and $|AB| = |CD| = |EF|=1$. Let G, H, and I be the midpoints of sides BC, DE, and FA respectively. Then GHI is an equilateral triangle.

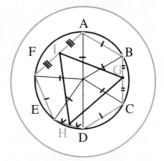

$\sqrt{-1}$ Complex Numbers

Food for Thought

❹. Equilateral triangles (with centers O_1, O_2, O_3) are constructed externally on the sides of a triangle ABC. Then $O_1O_2O_3$ is an equilateral triangle.

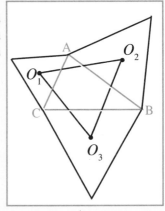

Napoleon's Theorem

Napoleon Bonaparte was fond of mathematics, especially geometry, and in 1794 he established the *École Polytechnique*, which produced most of the French mathematicians of the early nineteenth century. Although this theorem is credited to him, many historians are skeptical that Napoleon knew enough geometry to discover it.

*The use of complex numbers to solve three classical problems is discussed further in **Three Impossible Problems of Antiquity** (p. 215).*

Readers interested in learning more about the uses of complex numbers in geometry should look at Section 8.4 of *Problem-Solving Through Problems* by Loren C. Larson. For more information, see the *Annotated References*.

INFINITY REVISITED

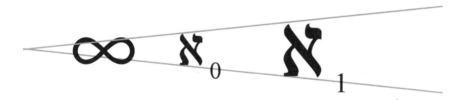

*Mathematical proofs, like diamonds, are
hard as well as clear, and will be touched
with nothing but strict reasoning.*

— John Locke

Paradox!

Humans have long been fascinated with the concept of infinity. The inability of language to express the infinite, however, has relegated the subject to the religious sphere for most of our history. Only since the time of the great German mathematician Georg Cantor (1845-1918) has some kind of mathematical study of infinity been undertaken, and the results have been as revelatory as they have been disturbing. In these few sections, we will introduce some ideas related to the mathematical notion of infinity.

The reader will likely have been introduced to infinity as some mysterious "number" that is bigger than all other numbers, and denoted by the symbol ∞. The concept of infinity is more complicated than it at first appears. Assumptions we take for granted with "finite" numbers no longer work. To see why, let's do some experimental mathematics. Consider the set of the positive integers:

$$\mathbb{N} = \{1, 2, 3,...\}$$

This is an infinitely large set, so let's call its cardinality ∞ . (The *cardinality* of a set is just its size. For example, the cardinality of the set {cat, dog, tree} is 3, because this set has 3 elements.)

Let's add "0" to the set $\mathbb{N}$. The resulting set is: $\{0, 1, 2, 3, ...\}$

In general, when we add an element to a set, we increase its cardinality by 1. (For example, if we add *house* to {*cat, dog, tree*}, we're left with {*cat, dog, tree, house*}, which has cardinality 4.) So the newly created infinite set has cardinality $\infty + 1$.

Next, add 1 to every element in this set. (This, of course, won't change the cardinality.) The newly created infinite set is:

$$\{1, 2, 3,...\}$$

Hold on a second! This is just our original set, which had cardinality ∞! So we've just proven that:

$$\infty = \infty + 1$$

Subtracting ∞ from each side of this equation, we get: $0 = 1$.

This can't be right! What's gone wrong here?

226

A classical paradox in philosophy tells of a (male) barber who lives in a small village. It is said that the barber shaves every man in the village who does not shave himself and *only* those who do not shave themeselves. The question is: *Does the barber shave himself?*

We can analyze this problem by assuming, in turn, each of the two possible answers and recording the conclusions in a table.

The barber shaves himself	The barber does not shave himself
Since the barber shaves only those who do not shave themselves, then the barber must be one of those who does not shave himself. This contradicts the statement that the barber shaves himself.	Since the barber shaves every man who does not shave himself, then the above statement requires that the barber shave himself. This contracts the statement that the barber does not shave himself.

Both statements lead to a contradiction. This paradox stems from the original condition which includes the barber in the set of people who are either shaved or not shaved by the barber.

Here is another paradox which resembles the quandry we encountered with the barber. Imagine a huge library, which contains not only every book ever written, but every book that ever could be written. Now imagine a book we'll call *The Ultimate Reference*, which is a list of all books **that do not contain their title in their text,** i.e. **which do not refer to themselves.** We now ask, "Is *The Ultimate Reference* listed in the *The Ultimate Reference?*" Think carefully; this is a much more complicated question than it first appears. As before, we assume each possible answer in turn.

The Ultimate Reference is listed	*The Ultimate Reference* is not listed
Since *The Ultimate Reference* lists only those books which do not refer to themselves, then the above statement implies that *The Ultimate Reference* does not refer to itself. This contradicts the statement given above.	Since *The Ultimate Reference* lists all those books which do not refer to themselves, then the above statement implies that *The Ultimate Reference* must refer to itself (i.e. list itself). This contradicts the statement given above.

Observe that each answer leads to a contradiction, and hence we have a paradox.

This example is known as *Russell's paradox* in honor of Bertrand Russell (1872-1970), a British mathematician, logician, philosopher, and peace activist. (Incidentally, he also won the Nobel Prize for Literature in 1950.) How can this paradox be resolved? Don't be disheartened if you can't figure it out immediately — the explosive debate caused by this issue sparked a revolution in set theory and logic in the early part of this century, and contributed to the development of a field devoted to the study of the foundations of mathematics.

**Bertrand Russell
1872-1970**

Notice that this paradox involved the idea of "self-reference". (A similar, but not so worrisome, paradox involving self-reference is the sentence "This sentence is false." Is it true? Is it false?) This is a recurring theme in modern mathematics, appearing for example in logic and in the study of fractals. In our future discussions about infinity, the flavor of this paradox will come up two or three more times — watch for it!

Food for Thought

You've now seen an example that suggests that $\infty = \infty + 1$. Can you think of an example that suggests that $\infty = \infty + \infty$?

Suggestions for Further Reading: There are a great number of popular books that discuss infinity at greater length, if not depth, than this book. A far more challenging but very rewarding read is Douglas Hofstadter's Pulitzer Prize-winning book, *Gödel, Escher, Bach: An Eternal Golden Braid.* For bibliographic information, see the *Annotated References.*

Infinity Revisited

Two Different Infinities

We discussed the concept of infinity in the previous section. This section will make our thinking about infinity more rigorous. We already have a vague idea of what infinity is, but in order to capture it in mathematical terms, we need a rigorous definition. Once we make this definition, we can prove many surprising facts (for example, that there are many different infinities).

To each set S, we will attach something called a *cardinality,* which roughly means its size. We'll use the notation $|S|$ to represent the cardinality of the set S. For example, the cardinality of {cat, dog, tree} is 3, and the cardinality of { } (the empty set) is zero, as it has no elements. We also have infinite sets, such as $\mathbb{Z}$ (the set of integers), $\mathbb{Q}$ (the rational numbers), and $\mathbb{R}$ (the real numbers).

We can compare the cardinality of finite sets very easily: $|\{cat,dog,tree\}| > |\{\}|$ because $3 > 0$. Furthermore, an infinite set is definitely bigger than any finite set. But it is far harder to compare the sizes of infinite sets. Are there more integers than points on a line? Are there more rational numbers than integers? (Your instinctive answer to this last question might be: the cardinality of the rational numbers is certainly greater than the cardinality of the integers, because every integer is a rational number, but not every rational number is an integer. But as you'll soon see, our definition of cardinality for infinite numbers will lead us to the conclusion that the set of rational numbers is the same size as the set of integers!)

How can we define cardinality for infinite sets? Imagine that you have a huge number of candies. Since you have a huge number of candies, you have a huge number of friends. How would you determine whether you have the same number of candies as friends? You could count both and compare, but there's another (more generous) way. You could start giving each of your friends a candy. If, at the end of the day, you've given each candy to a friend, and you've given each friend a candy, then you have the same number of candies as friends. (You'll likely be a little miffed, as there will be no candy left for you, but your altruism may sweeten your loss.)

The same process can be applied to compare cardinalities of infinite sets. Suppose we are given two sets S and T. If we can match up the elements of S and T so that each element of S is matched with a single element of T, and vice versa, then we will say that $|S| = |T|$. For example, the sets S = {1, 2, 3,...} and T = {0, 1, 2, 3, ...} are the same size, because we can match each element n of S with the element n-1 of T. (This was an example from the previous section. Notice that we can add an element to S, and we're left with a set of the same size. This is an unusual property of infinite sets.)

229

We now rephrase this example in more sophisticated language. Suppose we're given two sets S and T. A mapping from S to T is a function that associates to each element of S an element of T. This sounds complicated, but it's really very simple. The way to picture it is as follows. Draw two ovals on a piece of paper corresponding to the two sets. Write the elements of S in the first, and the elements of T in the second. This is a way of representing the two sets. Then a mapping can be thought of as this diagram, along with arrows pointing from the elements of S to the elements of T, where each element of S has exactly one arrow emanating from it.

For example, if S = {1,2,3,4} and T = {cat, dog, tree}, then the following is an example of a "mapping" from S to T.

Example of a Mapping

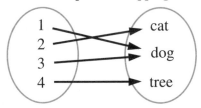

A *one-to-one* map is a mapping such that no two arrows point to the same element of T. The above map is not one-to-one. (Can you see why?) The following mapping *is* one-to-one.

Example of a One-to-One Mapping

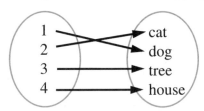

> If there is a one-to-one map from S to T, we say that *the cardinality of S is less than or equal to the cardinality of T.* We write |S| ≤ |T|.

Fortunately, this agrees with the usual meaning of the phrase, *less than or equal to* for finite sets. (It's always good to check that definitions agree with your intuitive sense of what's going on; this ensures that mathematicians are not trying to pull the wool over your eyes.)

We can use this definition for comparing cardinalities of infinite sets to establish a condition that two sets have equal cardinality.

> If there is a one-to-one map from S to T such that every element of T is hit by the map, then we say that *S and T are the same size.* We write |S| = |T|. We'll call this a *correspondence.*

(This is essentially the same way we checked that |set of candies| = |set of friends|.)

Now that we have a rigorous definition for determining when two infinite sets are the same size, we can show that there are infinite sets of different sizes. Let's go back to the first infinity we've met, |{ 1,2,3,...,}|. This infinity has a special symbol: $\aleph_0$. $\aleph$ (aleph) is the first letter of the Hebrew alphabet. $\aleph_0$, pronounced "alleff not", is the cardinality of the set of natural numbers, $\mathbb{N}$. A set of this size is said to be *countably infinite*, since being the same size as $\mathbb{N} = \{ 1, 2, 3,...\}$, it can be matched element-by-element with the positive integers.

In our search for an infinite set with greater cardinality than $\mathbb{N}$, we might consider the set of integers, $\mathbb{Z}$. It would seem that there are definitely more integers than there are positive integers. But when we apply the definition of cardinality given above, we discover that their cardinalities are the same! That is, $|\mathbb{N}| = |\mathbb{Z}|$ because there is a correspondence from $\mathbb{N}$ to $\mathbb{Z}$:

A correspondence from $\mathbb{N}$ to $\mathbb{Z}$

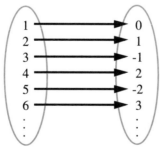

In this mapping, n is sent to $\dfrac{n}{2}$ if n is even, and $-\dfrac{(n-1)}{2}$ if n is odd. So instead of a new infinity, we've just stumbled on another version of our old one.

Cantor's Diagonalization Method

Now, we'll define a new infinity, $\mathbb{C}$, where $\mathbb{C}$ is defined to be the cardinality of the set of real numbers, $\mathbb{R}$. That is, $\mathbb{C} = |\mathbb{R}|$. Clearly, $\aleph_0 \leq \mathbb{C}$: there is an obvious one-to-one map from the positive integers to the real numbers. We'll soon show that $\mathbb{C} \neq \aleph_0$. This exceedingly clever proof was devised by Georg Cantor in 1874, and is known as the *Cantor diagonalization method*. It gives mathematical support to our intuition that there are somehow far more real numbers than positive integers. (And no, we can't simply say that the real numbers include the positive integers; just remember that $|\{1,2,3,...\}| = |\{2,3,4,...\}|$.)

Suppose we are given a mapping f from the integers $\mathbb{N}$ to the real numbers $\mathbb{R}$. This means that to each positive integer n, we can assign a real number $f(n)$. We'll see that some real number will be missed, so f can't be a correspondence.

Here is a possible start to our table for a purely arbitrary mapping f:
$$f(1) = 1.2$$
$$f(2) = \pi$$
$$f(3) = 203$$

First we make this into a big list, adding in some 0 digits.

$$...0\ 0\ 0\ 1\ .\ \textcircled{2}\ 0\ 0\ 0\ 0\ 0\ 0\ 0\ 0...$$
$$...0\ 0\ 0\ 3\ .\ 1\ \textcircled{4}\ 1\ 5\ 9\ 2\ 6\ 5\ 3...$$
$$...0\ 2\ 0\ 3\ .\ 0\ 0\ \textcircled{0}\ 0\ 0\ 0\ 0\ 0\ 0...$$

Then we circle the first digit after the decimal place in the first number, the second digit after the decimal place in the second number, and so on. Now, we make up a new real number, r, that lies between 0 and 1, according to the following rule.

> If the kth circled digit is not a 1, then the kth digit of r is 1.
> If the kth circled number is a 1, then the kth digit of r is 0.

Clearly r can't be the first number on the list, because it differs from the first number in (at the very least) the first digit after the decimal point. Similarly r can't be the second number on the list either, because it differs from the second number in (at least) the second digit after the decimal point. In fact, r can't be the nth number on the list, because it differs from the nth number in (at least) the nth digit

after the decimal point. So we've proven that r can't be on the list at all! In summary, we've proven that there is no correspondence from $\mathbb{N}$ to $\mathbb{R}$, so $\mathbb{C}$ is indeed bigger than $\aleph_0$!

Q. E. D. (Do you remember what Q.E.D. stands for?)

You might have noticed a similarity in flavor between this argument and Bertrand Russell's paradox in the previous section; this is an example of how mathematicians are able to turn bad news into good news.

If, after reading through the above proof, you're feeling overwhelmed, you may want to look over the proof a couple more times (or even sleep on it). When you've understood it, you should reach over your shoulder and pat yourself on your back: you've mastered some serious mathematics!

Food for Thought

❶. We've said that if there is a one-to-one mapping g from a set S to a set T, then $|S| \leq |T|$. You might imagine that if g is one-to-one, but misses some element of T, then $|S| < |T|$. Find an example to show that this is false. (Hint: You'll have to look at an infinite set. Another Hint: Re-read *Paradox,* p. 226.)

❷. Prove that the complex numbers are uncountable. In other words, show that $|\mathbb{C}| > \aleph_0$.

Are you ready for some more? In the next section, we'll see that there are in fact an infinite number of infinities. Or, if you'd like, you can now skip ahead to the section after that, where we'll use what we know about infinities to prove the existence of things called transcendental numbers.

An Infinitude of Infinities

For this section, you will need a good understanding of facts and ideas explored in the previous two sections.

Here is a proof, remarkable both for its elegance and its brevity, which shows that there are an infinite number of infinities of different sizes. We'll define infinite sets S_0, S_1, S_2, ... such that each is of a different cardinality.

Given a set S that is non-empty, define $\mathcal{P}(S)$ (called the *power set of* S) to be the set of subsets of S. So, for example, if S is {1,2,3}, then

$$\mathcal{P}(S) = \{\ \{\}, \{1\}, \{2\}, \{3\}, \{1,2\}, \{1,3\}, \{2,3\}, \{1,2,3\}\ \}.$$

Note that $|S| = 3$ and $|\mathcal{P}(S)| = 8$. (It is no coincidence that $|\mathcal{P}(S)| = 2^{|S|}$, and those readers with some experience with combinatorics, the study of counting, might be able to see why.) If S is a finite set, it's clear that $|S| < |\mathcal{P}(S)|$. Remarkably, this result actually holds for infinite sets, too.

Before we actually go through the proof, let's see what this means. We have one infinity, $\aleph_0 = |\mathbb{N}|$. This proof shows the existence of a sequence $|\mathbb{N}|$, $|\mathcal{P}(\mathbb{N})|$, $|\mathcal{P}(\mathcal{P}(\mathbb{N}))|$, $|\mathcal{P}(\mathcal{P}(\mathcal{P}(\mathbb{N})))|$,..., where each term is less than the one following it. So we have an increasing sequence of infinities — countably infinite infinities!

Theorem $|S| < |\mathcal{P}(S)|$

Proof.

Consider the one-to-one function, f, which maps each element s of S to the singleton set {s} of $\mathcal{P}(S)$. Since $f: S \longrightarrow \mathcal{P}(S)$, given by $f(s) = \{s\}$, is a one-to-one function, $|S| \leq |\mathcal{P}(S)|$. (Here we are using a fact which appeared in *Two Different Infinities*, p. 229.) So we have only to prove that $|S| \neq |\mathcal{P}(S)|$. If we can show that every mapping from S to $\mathcal{P}(S)$ misses an element of $\mathcal{P}(S)$, which we'll soon do, then there can't be a correspondence from S to $\mathcal{P}(S)$, so $|S| \neq |\mathcal{P}(S)|$. Putting these two facts together, $|S| \leq |\mathcal{P}(S)|$ and $|S| \neq |\mathcal{P}(S)|$, so $|S| < |\mathcal{P}(S)|$.

Let g be any mapping from S to $\mathcal{P}(S)$. This means that to each element of S, g assigns a particular subset of S. Now let's show that there is an element of $\mathcal{P}(S)$ missed by g. Consider the set $T = \{s \in S \mid s$ is not in $g(s)\}$. (This is clearly a subset of S since all of its elements are taken from S.)

234

A One-to-One Mapping g of a Set S to its Power Set $\mathcal{P}$ (S)

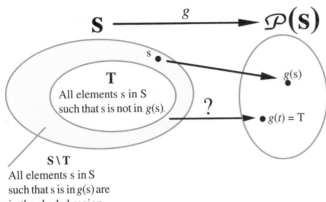

$$S \xrightarrow{\quad g \quad} \mathcal{P}(S)$$

T
All elements s in S
such that s is not in g(s).

?

g(s)

g(t) = T

S \ T
All elements s in S
such that s is in g(s) are
in the shaded region

Here is the crux of the argument: *Is there an element t of S so that g(t)=T?*
As in the discussion of Russell's paradox (p. 227), we tabulate the two possible answers.

Answer 1: $g(t)$ = T and t belongs to T	Answer 2: $g(t)$ = T and t does not belong to T
Since t is in g(t), then (by definition of T) t is not in T. This contradicts Answer 1 above, which said that t belongs to T.	Since t is not in g(t), then (by definition of T) t is in T. This contradicts Answer 2 above, which said that t does not belong to T.

That is, if $g(t)$ = T, then both statements, t in T and t not in T are false. No matter how we squirm, we can't escape this contradiction. We couldn't have had this t in the first place, so T was definitely missed by the mapping g. This was all we needed to complete the proof.

Now that the proof is done, let's look at what makes it work. We showed that given any mapping we had from S to $\mathcal{P}$(S), we could always construct an element of $\mathcal{P}$(S) missed by g. This showed that no correspondence from S to $\mathcal{P}$(S) existed, so $|S| \neq |\mathcal{P}(S)|$. Furthermore, $|S| \leq |\mathcal{P}(S)|$, so we get our desired result: $|S| < |\mathcal{P}(S)|$.

*To see how to use infinities to show that something **does** exist, turn ahead to*
The Existence of Transcendental Numbers (p. 237).

Historical Digression

The Man Who Shook the Foundations of Mathematics

By 1874, mathematicians such as Augustin-Louis Cauchy, Karl Weierstrass, and Gauss had succeeded in formalizing most of the intuitive concepts in mathematics to provide more rigorous methods of proof. For example, the famous ε–δ definition of continuity was introduced as an algebraically constructive way to test whether a function is continuous. It was generally believed that all of mathematics would ultimately be derivable from a finite set of consistent axioms. During this era of naïve optimism, Georg Cantor, a German mathematician published his revolutionary paper on *Mengenlehre* — a theory of infinite sets or classes. Cantor's profound paper

Georg Cantor

1845 - 1918

of 1874 established a method for assigning cardinalities to infinite sets. Through the construction of a correspondence between the set of integers, ℤ, and the set of rational numbers, ℚ, he was able to show that both these sets have the same cardinality even though ℤ is a proper subset of ℚ! Furthermore, Cantor's paper showed that the set of algebraic numbers, 𝔸, has the same cardinality as both ℤ and ℚ! (In a few pages, we will see this argument.) This result was seen as a paradox because any proper subset of a finite set has smaller cardinality.

Cantor's papers on *Mengenlehre* were met with a barrage of criticism by the conservative mathematicians of the day. Among the most vitriolic opponents of his approach to infinite sets was Leopold Kronecker, who challenged Cantor's use of indirect proof to establish the existence of certain sets . The negative reaction to his work and the lack of acceptance by the mathematicians in the mainstream of mathematics caused Cantor to suffer extreme bouts of depression and self-doubt. The controversy raged on and finally in 1884, at age 40, Cantor suffered a complete mental breakdown. He recovered partially from this breakdown and continued to make important contributions to his *Mengenlehre* between subsequent recurrences of depression. By 1891, Cantor's ideas were gaining acceptance in the mathematics community and Cantor had reconciled his differences with Kronecker. On January 6, 1918, Cantor died in a mental hospital in Halle, Germany. Today, Cantor's work stands as one of the great turning points in mathematics, marking the end of an innocent belief in unlimited rigor and the beginning of a more introspective look at mathematical truth.

The Existence of Transcendental Numbers

In the first two sections of this chapter, we developed the tools needed to discuss infinities. It is now time to put these tools to good use in another field of mathematics. First, we need a definition.

Definition. An *algebraic number* is any complex number that satisfies a polynomial equation,

$$a_n x^n + a_{n-1} x^{n-1} + \ldots + a_1 x + a_0 = 0$$

where the co-efficients, a_n, a_{n-1}, ..., a_1, a_0 are all integers.

For example, $3/2$, 0, and $\sqrt[5]{2}$ are all algebraic. Can you find the polynomial equations which have them as roots? Can you see that every rational number is algebraic? If we designate the set of algebraic numbers by $\mathbb{A}$, then, in set-theoretic notation, we write $\mathbb{Q} \subset \mathbb{A}$. That is, adding all the irrational algebraic numbers to the set of rational numbers expands our number set to include many more complex numbers. But do we get them all? Is $\mathbb{A} = \mathbb{C}$? Are there any complex numbers that are *not* algebraic? Such numbers are called *transcendental numbers* (described in *Numbers, Numbers, and More Numbers*, p. 118). Using our mathematics of the infinite, we will show that transcendental numbers actually do exist, and that in some sense "most" complex numbers are transcendental.

It is true that π and e are transcendental, but the proofs of these facts are quite difficult. It is easier (but still hard) to show that the number

$$\sum_{n=1}^{\infty} \frac{1}{10^{n!}}$$

is transcendental. In general, it is very tricky to show that any particular real number is transcendental. So we're in the strange position of knowing that these transcendental numbers are almost everywhere on the real line without being able to point at one and say, "That's transcendental."

Transcendental Meditation
—Mathematician-Style

π, e, ...

The existence of the transcendental numbers derives from the fact that the set of complex numbers is much larger than the set of algebraic numbers. This in turn hinges on the fact that the algebraic numbers are countable while the complex numbers are not. By this, we mean that $|\mathbb{A}| = |\mathbb{N}|$ (soon to be proved) and $|\mathbb{N}| < |\mathbb{C}|$ (shown in *Two Different Infinities*, p. 229). So this will imply that $|\mathbb{A}| < |\mathbb{C}|$, i.e. there are many complex numbers which are not algebraic.

Subtle Theorem: $|\mathbb{A}| = |\mathbb{N}|$

Proof. At first glance, this statement seems highly improbable. There are so many more algebraic numbers than natural numbers! But remember that we are thinking in terms of cardinality. We want a correspondence between the natural numbers and the algebraic numbers. In other words, to every natural number n, we want to assign an algebraic number $f(n)$, so that every algebraic number is $f(n)$ for some n. Here is a method for making that big list.

This proof is not simple. But it is worthwhile!

Each algebraic number x is a root of a particular polynomial equation of degree m, for some m. Furthermore, all coefficients of that polynomial equation have absolute value less than or equal to a for some integer a. Therefore when we list the roots of all polynomial equations of degree m and with coefficients a_i such that $|a_i| \le a$, then x will appear in that list. It remains only to show how we can list all the polynomial equations of degree m and with coefficients a_i such that $|a_i| \le a$ for all positive integral values of m and a. The diagram shows the famous diagonal process which creates such a list by associating the ordered pair (m, a) with the set of polynomial equations of degree m and coefficient upper bound a.

upper bound, a, for the largest absolute value of the coefficients in the polynomial equation $\longrightarrow$

degree of the polynomial equation $\downarrow$

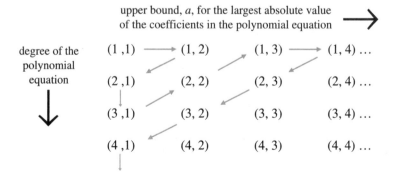

As the diagram suggests, we list first all the algebraic numbers that are the roots of some polynomial equation of degree 1, all of whose coefficients have absolute value at most 1. There are three such numbers: -1, 0, and 1. (These are the respective roots of the polynomial equations: $x + 1 = 0$, $x = 0$, and $x - 1 = 0$.) Let these be the first three values of f: $f(1) = -1$, $f(2) = 0$, and $f(3) = 1$.

Next we list all algebraic numbers *not already listed* that are the solutions to some polynomial of degree 1, all of whose coefficients have absolute value at most 2. There are a finite number: -2, -1/2, 1/2, and 2. (These are the respective roots of $x + 2 = 0$, $2x + 1 = 0$, $2x - 1 = 0$, and $x - 2 = 0$.) Let these be the next values of f: $f(4) = -2$, $f(5) = -1/2$, $f(6) = 1/2$, and $f(7) = 2$.

Next, list all algebraic numbers not already listed that are solutions to some polynomial of degree 2, all of whose coefficients have absolute value at most 1. (There are a finite number.) Let these be the next values of f.

Next, do the same thing with all new algebraic numbers that are roots of a polynomial of degree 3 with co-efficients of absolute value at most 1. We proceed in this manner following the sequence shown by the arrows; that is,

> degree 2 absolute value ≤ 2, then
> degree 1 absolute value ≤ 3, then
> degree 1 absolute value ≤ 4, then
> degree 2 absolute value ≤ 3, then
> degree 3 absolute value ≤ 2, then
> degree 4 absolute value ≤ 1, then
> degree 5 absolute value ≤ 1...

And so on.

At each stage, we get a finite number of new algebraic numbers (as there are only a finite number of polynomials of degree m and with coefficients of absolute value at most a; in fact, at most $(2a+1)^{(m+1)}$ of them).

Every algebraic number will eventually be generated in this fashion. For example, if α is a root of $14x^{304} - 300x^{302} + 4x + 1 = 0$, it will be on our list; we'll eventually get to polynomials of degree 304 and coefficients of absolute value at most 300.

As every algebraic number will appear on this list $f(1)$, $f(2)$, $f(3)$, ... exactly once, we have described a correspondence between $\mathbb{N}$ and $\mathbb{A}$, so $|\mathbb{N}| = |\mathbb{A}|$.

In summary, we find the following surprising situation. There are seemingly far more algebraic numbers than integers. While algebraic numbers seem ubiquitous, it is difficult to find a transcendental number and *prove* that it is transcendental. Yet, in terms of cardinality, there are no more algebraic numbers than integers, and there are far more transcendental numbers than algebraic numbers.

Food for Thought

(as if there weren't enough here already!)

❶. Can you show that $|\mathbb{N}| = |\mathbb{Q}|$?

❷. Can you prove the existence of a *real* transcendental number?

❸. At first it may seem like the following would be an easier way to get a correspondence between $\mathbb{N}$ and $\mathbb{A}$:

i) List all the algebraic numbers that are the zeroes of a polynomial of degree 1.
ii) List all the algebraic numbers (not already listed) that are the zeroes of a polynomial of degree 2.
iii) List all the algebraic numbers (not already listed) that are the zeroes of a polynomial of degree 3.
...And so on.

But with this method, we would never reach step ii): there are too many algebraic numbers that are the zeroes of degree 1 polynomials. What is going wrong here? How did our original proof get around this problem?

❹. a) You can think about how to prove the following result. Let S_1, S_2, S_3, ... be finite sets that get larger and larger in size. Create a correspondence to prove that:

$$\mathbb{N} = \left| \bigcup_{j=1}^{\infty} S_j \right|$$

b) Now let $S_j = \{$those algebraic numbers that are solutions of a polynomial of degree m, whose co-efficients have absolute value at most a, where $m+a = j\}$. Prove that S_j is a finite set. Prove that every algebraic number appears in some S_j. Then use the result of part a) to show that $|\mathbb{N}| = |\mathbb{A}|$. (This is essentially the same proof as the one given in the text, although it is conceptually a little cleaner.)

240

The Youngest Tenured Professor in Harvard History

Noam Elkies (USA)
Born August 25, 1966.

Noam Elkies has been playing around with arithmetic and number puzzles for as long as he can remember. In early childhood he enjoyed counting the keys on a piano, an activity that would foreshadow his adult fascination with both mathematics and music. Thanks to special camps and enriched education in Israel between 1970 and 1978, his ability in mathematics was encouraged from an early age. He happened across a Hebrew translation of Euclid's *Elements* on his parents' bookshelves, and fell in love with mathematics from reading it.

When Noam returned to New York, he enroled in Stuyvesant High School, perhaps the pre-eminent mathematical school in the U.S. He plundered the public libraries in search of more mathematics. Martin Gardner's books were a particular source of inspiration. Indeed, one of Gardner's problems later became a prize-winning project for Noam at the 1982 *Westinghouse Science Talent Search*, a national science fair.

Noam also competed in the series of competitions run by the Mathematical Association of America. The pinnacle of this series is the *USA Mathematical Olympiad*, which Noam won in his final two years of high school. Based on that performance, he was invited to compete for the USA at the *International Mathematical Olympiad*, where he won Gold Medals in each of his two years.

While at Stuyvesant, Noam also attended the Juilliard school of music. Noam considers mathematics and music similar aesthetic pursuits; abstraction and a certain playfulness are essential to both. In fact, Noam considered a career in music as late as his final year of high school; even today he gives regular recitals, sometimes of his own compositions.

After Stuyvesant and Juilliard, Noam studied mathematics and music at Columbia University, graduating after three years when he was only 18. In each of those three years, he won top honors in the North American *Putnam Mathematical Competition*.

In 1985, Noam entered the Ph.D. program at Harvard. In the summer after his first year, he stunned his mentors by proving a conjecture in number theory that had defied mathematicians for over 25 years. This important result became his Ph.D. thesis, and the following year, after only two years of study, he received his doctorate in mathematics.

In 1987, another long-standing problem succumbed to Noam's charms. It had been known for centuries that it is not possible for one cube to be written as the sum of two cubes. But sometimes a cube can be written as the sum of *three* cubes; for example, $3^3 + 4^3 + 5^3 = 6^3$. Leonhard Euler (1707-1783) conjectured that for $n > 2$ it is possible for n n^{th} powers to add to another n^{th} power, but impossible for n-1 to do so. In 1966, L.J. Lander and T.R. Parkin showed that Euler's conjecture is false: $27^5 + 84^5 + 110^5 + 133^5 = 144^5$. But it had been unknown since Euler's day whether there are three fourth powers summing to a fourth. Using elegant and sophisticated tools of number theory, Noam hunted down a counterexample:

$$2682440^4 + 15365639^4 + 18796760^4 = 20615673^4$$

He then proved that, in some sense, there are *lots* of counterexamples, and showed how to produce an infinite number of them. (The counterexamples are very large; the numbers in Noam's second counterexample have almost seventy digits each!)

Noam's main concern about graduating early was that he would be forced to leave Harvard's active academic environment too soon. Such fears would prove groundless. He was offered a Junior Fellowship from 1987-90, and then an Associate Professorship from 1990-93. Finally, in 1993, at the age of 26, he became the youngest person ever to be granted tenure at Harvard University.

$\mathcal{P}$ersonal $\mathcal{P}$rofile

Noam's favorite problems include the famous *Riemann Hypothesis*, a long-standing conjecture in number theory that has tremendous ramifications in the field. "All number theorists are inherently fascinated by it," says Noam. But despite the concentrated effort being expended on this one problem, Noam notes that, "it might well be solved by accident."

Another problem Noam has attacked is the *ABC Conjecture* which lies at the crossroads between number theory and geometry. The ABC Conjecture implies many important results, including Fermat's Last Theorem. Noam proved another important implication of this conjecture in 1991.

The ABC Conjecture (due to Masser and Oesterlé) is reasonably simple to understand. Imagine that there were a solution to Fermat's Last Theorem:

$$x^n + y^n = z^n \quad (n > 3).$$

Let $A = x^n$, $B = y^n$, and $C = z^n$, so $A + B = C$. Then the product of the primes dividing ABC is at most $xyz < C^{3/n}$, which is much less than C. If we could show that this is impossible (when phrased more rigorously) then we could prove Fermat's Last Theorem.

The ABC Conjecture is basically just that: if A, B, and C are relatively prime, and $A + B = C$, then the product of primes dividing ABC should be much greater than $C^{1-\varepsilon}$. Put very loosely, if a little carelessly, in examples when C is very large, the product must be at least of the same order of magnitude as C. The ABC Conjecture, altered slightly, is known to be true if A, B, and C are polynomials instead of integers. The proof of this result is very similar to the solution to the following problem from the *Sixteenth William Lowell Putnam Mathematical Competition* in 1956:

> The polynomials $P(z)$ and $Q(z)$ with complex coefficients have the same set of numbers for their zeros but possibly different multiplicities. The same is true of the polynomials $P(z) + 1$ and $Q(z) + 1$. Prove that $P(z) \equiv Q(z)$.

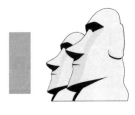
Noam has also suggested another unsolved problem that you might wish to investigate. This problem is a more general form of the following question first formulated by Sam Loyd (1841-1911), a renowned American puzzlist.

What is the greatest number of line segments which can be drawn through 16 points so that each line segment contains 4 points?

In the diagram below, there are sixteen dots and ten line segments, with four dots on each line segment. We will write this as "16 = 10 ⊗ 4". (Here, the symbols = and ⊗ do not have their usual meanings.)

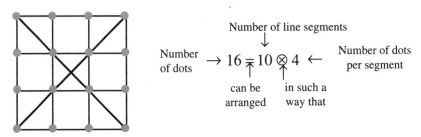

However, in the following diagram, there are sixteen dots and fifteen lines, with four dots on each line. (We write this as "16 = 15 ⊗ 4".) This is the best-known result for 16 dots and 4 dots per line segment.

16 = 15 ⊗ 4

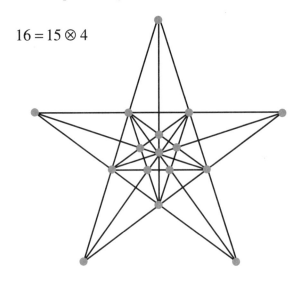

This case can be generalized to get $n^2 = (3n+3) \otimes n$, where n is even (and at least 4). Can you see how? Try it with $n = 6$. Is this the best that can be done? The following diagram shows the best result for $n = 3$:

$$9 = 10 \otimes 3$$

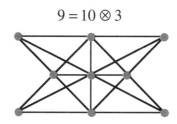

In 1981, Noam discovered the following example that is the best known for $n = 5$:

$$25 = 18 \otimes 5$$

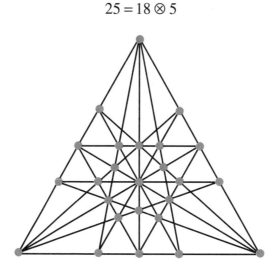

It isn't even known if $n^2 = (3n + 3) \otimes n$ is possible for n odd and greater than 5. Can you find an example that shows that $7^2 = 24 \otimes 7$? How close can you get? Also, can you prove that one can never get $n^2 = m \otimes n$ for $m > 3n+3$?

Afterword

I hope you have enjoyed these forays into mathematics as much as I have, and that you have been left with much to think about. The moral to this story is that beautiful, surprising results are always worthwhile for their own sake, but they can often lead into deeper and even more exciting waters.

While a student at Northern Secondary School in Toronto, Kevin Purbhoo came up with the following beautiful problem:

> On a remote Norwegian mountain top, there is a huge checkerboard, 1000 squares wide and 1000 squares long, surrounded by steep cliffs to the north, south, east, and west. Each square is marked with an arrow pointing in one of the eight compass directions, so (with the possible exception of some squares on the edges) each square has an arrow pointing to one of its eight nearest neighbors. The arrows on squares sharing an edge differ by at most 45°. A lemming is placed randomly on one of the squares, and it jumps from square to square following the arrows.
>
> Prove that the poor creature will eventually plunge from a cliff to its death.

Although Kevin did not know it at the time, this problem anticipates several subtle and important results in topology, a type of geometric thinking useful in most areas of mathematics. This is only one example where a simple idea or problem opens up new mathematical horizons.

If you are a high school student and want to learn more interesting mathematics, there are many avenues you can pursue. First of all, become comfortable writing proofs. Putting ideas to paper in a form convincing to others sharpens the mind; like any new skill, it requires practice. Euclidean geometry provides a perfect opportunity for this, as the tools required are minimal and the rewards great. (Sadly, geometry seems to be quietly disappearing from the high school curriculum.)

The beauty of mathematics has inspired many authors to write expository books. As a result, there are numerous excellent mathematics books aimed at many levels. The *Annotated References* is a good starting point, but there are many other worthwhile books not listed. Browse around, and follow up on authors and subjects you really enjoy. If your school library has only a few titles, convince your librarian to buy more.

Although private reading and independent study are useful, it is important to share your ideas with teachers and other students. A Math Club is a one way to form such a community; arguing about ideas with friends is another. There are also many summer and weekend programs aimed at students; your local university is a good source.

Finally, you might enjoy taking part in mathematical competitions. In Canada, the University of Waterloo runs an excellent series of competitions at many grade levels. (For more information, contact the Canadian Mathematics Competition, Faculty of Mathematics, University of Waterloo, Waterloo, Ontario, Canada N2L 3G1, tel. (519) 885-1211 ext. 2248, fax (519) 746-6592.) Success here opens the door to many other competitions, including the *Canadian Mathematical Olympiad* and the *International Mathematical Olympiad*. In both Canada and the U.S., the Mathematical Association of America runs a cycle of competitions beginning with the *American High School Mathematical Examination* (contact: Dr. Walter E. Mientka, Executive Director, American Mathematics Competition, 1740 Vine Street, University of Nebraska, Lincoln NE, 68588-0658 USA). Teams from parts of the U.S. and Canada compete every May at the fun and informal *American Regions Mathematics League.* (contact: Mr. Mark E. Saul, ARML President, Bronxville School, Bronxville NY, 10708 USA, (914) 337-5600, e-mail 73047.3156 @compuserve.com). Finally, the *International Mathematics Talent Search* (contact: Dr. George Berzsenyi, Department of Mathematics / Box 121, Rose-Hulman Institute of Technology, Terre Haute IN 47803-3999 USA) is a non-competitive year-long problem-solving program. While most other competitions have stringent time limits, the IMTS allows more reflection on the part of the participants, fostering not only ingenuity, fast thinking, and creativity, but also commitment, reliability, and perseverance.

Once again, I hope you have been challenged and excited by this book. If you have any comments, please write. Best of luck in future explorations!

Ravi Vakil
Department of Mathematics
Harvard University
One Oxford St.
Cambridge, MA
USA 02138-2901

Annotated References

Berlekamp, Elwyn R., Conway, John H., and Guy, Richard K. *Winning Ways for your Mathematical Plays.* New York: Academic Press, 1982.
These two volumes explain mathematical game theory with many examples of fun games to play and win. Well-written but intellectually demanding.

Boyer, Carl B. and Merzbach, Uta C. *A History of Mathematics.* 2nd ed. New York: John Wiley & Sons, 1991.

Ball, W.W. Rouse and Coxeter, H.S.M. *Mathematical Recreations and Essays.* 13th ed. New York: Dover Publications Inc., 1987.

Beckmann, Petr. *A History of Pi.* Boulder, Colorado: The Golem Press, 1971.

Castellanos, Dario. "The Ubiquitous π". *Mathematics Magazine.* Vol. 61, No. 22, Apr. 1988, 67 - 98.

Coxeter, H.S.M. "The Golden Section, Phyllotaxis, and Wythoff's Game", *Scripta Mathematica.* Vol. 19, No. 2-3, June-Sept. 1953, 135 - 43.

*Coxeter, H.S.M. and Greitzer, Samuel L. *Geometry Revisited.* Washington D.C.: Mathematical Association of America (MAA), 1967.
This is Volume 19 in the MAA's excellent *New Mathematical Library.* For more information about any of the MAA's publications, call 1- 800-331-1622 or (202) 387-5200.

Crux Mathematicorum.
Crux is a high-level problem-solving journal published by the Canadian Mathematical Society; it appears ten times per year, and includes a regular "Olympiad Corner". For more information, contact: CMS/SMC, 577 King Edward, Suite 109, POB/CB 450, Station A, Ottawa, Ontario, Canada K1N 6N5, tel. (613) 564-2223, fax (613) 565-1539.

Curl, Robert F. and Smalley, Richard E. "Fullerenes", *Scientific American.* Vol. 264, No. 4, Oct. 1991, 54-63.

Gardner, Martin. "Six sensational discoveries that somehow or another have escaped public attention". *Scientific American* Vol. 232, No. 4, Apr. 1975, 126-133.
Keep an eye out for anything Gardner has written!

Annotated References (cont'd)

Gleick, James. *Genius.* New York: Vintage Books, 1992.

*Greitzer, Samuel L., *Arbelos.* Washington D.C.: MAA, 1982-1988. The *Arbelos* was a first-rate journal of problem-solving mathematics for high school students that was essentially the work of one man.

Hardy, G. H. *A Mathematician's Apology.* Cambridge: Cambridge University Press, 1993.

Hoffman, Paul. *Archimedes' Revenge: The Joys and Perils of Mathematics.* New York: W.W. Norton & Co., 1988.

Hofstadter, Douglas R. *Gödel, Escher, Bach: An Eternal Golden Braid.* New York: Random House, 1979.
Brilliant; winner of the Pulitzer Prize.

Holton, Derek. *Let's Solve Some Math Problems.* Waterloo, Ontario: Waterloo Mathematics Foundation, 1993.
For more information, contact the Canadian Mathematics Competition, Faculty of Mathematics, University of Waterloo, Waterloo, Ontario, Canada N2L 3G1, tel. (519) 885-1211 ext. 2248, fax (519) 746-6592.

*Honsberger, Ross. *Ingenuity in Mathematics.* Washington D.C.: MAA, 1970.
Ingenuity is another book from the MAA's *New Mathematical Library.* The MAA also has other titles by Honsberger, all of them excellent.

Jacobs, Harold R. *Mathematics, A Human Endeavor.* San Francisco: W.H. Freeman, 1982.
A teacher's guide is also available.

*Larson, Loren C. *Problem-Solving Through Problems.* New York: Springer-Verlag, 1983.
This is an excellent book in general. Although it is intended for undergraduates, many parts are accessible (and interesting) to gifted high school students. Chapters include: Induction and Pigeonhole, Arithmetic, Algebra, Complex Numbers, Inequalities, and Geometry. Call Springer-Verlag at 1-800-777-4643 for more information.

Annotated References (cont'd)

Mathematical Digest.

Mathematical Digest is an excellent journal for high school students appearing four times per year, edited by Dr. John Webb at the University of Cape Town in South Africa. For more information, write to Mathematical Digest, Department of Mathematics, University of Cape Town, 7700 Rondebosch, South Africa.

*Mathematical Mayhem.

Mayhem is a non-profit problem-solving journal appearing five times per year, written by and for high school and university students. There are informative and irreverent articles as well as problems at different levels. For more information, contact the Department of Mathematics, University of Toronto, 100 St. George St., Toronto, Ontario, M5S 1A1, or e-mail mayhem@math.toronto.edu.

Nelsen, Roger B. *Proofs Without Words: Exercises in Visual Thinking.* Washington D.C.: MAA, 1993.

*Polya, G. *How to Solve It.* Princeton: Princeton University Press, 1973.

Rucker, Rudy. Infinity and the Mind: *The Science and Philosophy of the Infinite.* Boston: Birkhauser, 1982.

Swetz, Frank J. ed. *From Five Fingers to Infinity: A Journey through the History of Mathematics.* Chicago: Open Court Publishing Company, 1994.

*I strongly believe that every high school should have these books on hand.

Index

Index (cont'd)

Index (cont'd)

ORDER FORM

Fax orders: (905) 335 - 5104

Telephone orders: (905) 335 - 5954

Postal Orders: Brendan Kelly Publishing Inc.
2122 Highview Drive
Burlington, Ontario, CANADA
L7R 3X4

Please ship ☐ **additional copies of this book to the following name and address:**

_____

_____

_____

_____

_____

If ordering by mail, please enclose a cheque payable to **Brendan Kelly Publishing Inc.** for $16.95 in **U.S. funds per copy**. Please add $4.00 in U. S. funds to the total cost to cover postage and handling. (In mathematical language, the total payment remitted in U.S. dollars should be:

$16.95n + 4$, where n is the number of books ordered.)

If ordering by fax, you will receive an invoice for the amount calculated above.

Discounts are given for large orders. Please call for details.